Making
Physics
FUN

Dedication

I have been teaching college-level physics for nearly three decades. Over that time I have continually improved my college teaching practice by observing and working with many talented elementary and middle school science teachers.
This book is dedicated to all of you.

Making
Physics
FUN

Key Concepts,
Classroom Activities,
& Everyday Examples,
Grades K–8

Robert Prigo

Skyhorse Publishing

Skyhorse Publishing books may be purchased in bulk at special discounts for sales promotion, corporate gifts, fund-raising, or educational purposes. Special editions can also be created to specifications. For details, contact the Special Sales Department, Skyhorse Publishing, 307 West 36th Street, 11th Floor, New York, NY 10018 or info@skyhorsepublishing.com

Skyhorse® and Skyhorse Publishing® are registered trademarks of Skyhorse Publishing, Inc.®, a Delaware corporation.

Visit our website at www.skyhorsepublishing.com.

10 9 8 7 6 5 4 3 2 1

Library of Congress Cataloging-in-Publication Data is available on file.

Cover design by Lisa Miller

Print ISBN: 978-1-62914-744-4
EBook ISBN: 978-1-63220-037-2

Printed in the United States of America

Contents

Preface

This book is primarily for elementary and middle school inservice and preservice teachers who are already familiar with constructivist methods in science education, including inquiry-based teaching strategies, formative assessment techniques, and the design of lesson plans consistent with state and national science standards. While such teaching methods are not addressed in this book, at least not directly, familiarity with such methods would definitely complement the more practical focus of this book. This is a nuts-and-bolts resource book for the elementary or middle school teacher who is looking to design or is in the process of designing lesson plans or units in specific areas in physical science. This book would also be a valuable resource for home-schooling parents or guardians who are looking to augment or develop science lessons at home with a greater emphasis on hands-on activities and relevant examples from the everyday world. This book may also be of value to the high school physics teacher who is looking to extend his or her repertoire of classroom demonstrations and real-world examples in topic areas often taught at the secondary level.

Each chapter presents succinct descriptions of the physical science concepts. This narrative is followed by a description of many engaging activities that illustrate those concepts. This in turn is followed by a list of interesting everyday examples that support and complement those activities. I feel that the activities and everyday examples are essential to constructing an understanding of the concepts. Each chapter ends with a circus of activities that could be used to initiate the unit and some sample investigable questions that could be used to begin a full cycle of inquiry.

This book is not a set of lesson plans, nor does this book give any kind of prescriptive, day-to-day teaching outline. The unit design and teaching methodology, as they should be, are in the capable hands of the classroom teacher. This was done purposefully so the classroom teacher would not feel constrained by a single delivery system and could choose a variety of uses for the book's contents within his or her teaching style and context.

This book addresses five specific areas of physical science (Motion and Force, Fluids and Buoyancy, Waves and Sound, Light and Electromagnetic Waves, and Electricity and Magnetism) with dozens of suggested classroom activities, an extensive list of everyday examples, some ideas for initiating the unit, and sample inquiry questions—all of which can be mixed and matched to form a complete unit.

This book has emerged in response to the specific needs of classroom teachers and method students who are faced with the challenging task of putting together a physical science unit. Indeed, it is an outgrowth of two decades of professional development work with hundreds of preservice teachers, classroom teachers, and school administrators. From the beginning, these professional development efforts, often supported by major grants from the National Science Foundation, focused on the "big ideas" in K–8 science education. We articulated and modeled an inquiry-based pedagogy (Harlen, 2001; Saul & Reardon, 1996) to support the development of physical science concepts, science process skills, and the dispositions of science. We developed and practiced innovative formative assessment techniques (Carlson, Humphrey, & Reinhardt, 2003) that were consistent with the inquiry approach and concept formation. And we revealed the many ways the *National Science Education Standards* (National Research Council, 1999) supported our focus on science concept development, authentic inquiry, and formative assessment. But this was not enough. At the same time teachers and future teachers were reading about, practicing, and getting excited about the big ideas in K–8 science teaching, learning, and assessment, they were also facing the more down-to-earth and always difficult task of actually designing science units and lesson plans for their classrooms. This book is more about the latter and less about the former.

To be more specific, this book is the answer to the following questions that arise when the actual task of planning a unit begins:

- Where should I go to review (or learn) the important physical concepts (content) that should be addressed within this topic?
- Where should I go to find a variety of appropriate and engaging hands-on activities that can help reveal those important physical concepts?
- Where should I go to find some everyday examples of the concepts in the natural and person-made world around me to help elucidate the physical concepts?
- Where should I go to find an activity or activities that I might use to invite my students into the unit (an invitation to learn)?

- Where should I go to find some sample investigable questions that students might pursue when participating in the cycle of inquiry?

The organization of this book follows directly from the five content areas. For each area, a synopsis is given of the fundamental physical concepts. The synopsis is in narrative form and nonmathematical, with emphasis being given only to those concepts that are fundamental to the topic area. It is hoped that when this narrative is combined with the many activities and everyday examples, the full meaning and explanatory power of the concepts will become apparent. The pedagogical strategy of integrating concept description with multiple activities and numerous everyday examples can lead to a deeper understanding of the concepts. Indeed, the delight that comes from "seeing the unity" that the concepts provide by explaining diverse experiences and everyday examples can be intellectually stimulating and make the learner eager for more (Prigo, 1994).

The list and description of each of the activities will be limited to those activities that satisfy the following criteria:

The materials needed to set up the activity are relatively easy to obtain. In those cases where specific items may be more difficult to obtain, suggestions will be given for an appropriate vendor.

The concept is clearly displayed in the activity. An activity that may confuse or display outcomes that hinder insight into the particular concept is not included.

The activity can be made developmentally appropriate. Adjustments will be needed for various grade levels, but most of the activities can be made appropriate for elementary and middle school classrooms.

The activity is inviting and engaging. The students will be attracted to the activity and readily engage with the materials.

The activity is ripe for further investigation and inquiry. When manipulating the materials, new ideas and further questions will come to mind and lead to further investigations.

Following the activities section, the "Everyday Examples" section presents a laundry list of various situations where the concept is revealed in the world around us. These examples are taken from the workings of the natural world as well as from specific human applications. Combining these everyday examples with the hands-on activities helps to authenticate the concept and reveal its power by unifying diverse phenomena and experiences. While most of these examples come from processes in the natural world around us, many will also be cited from the worlds of sports, medicine, engineering, the backyard, and the habits of our daily lives.

A wonderful way to invite students into a new science unit is through what I fondly call the "concept circus" format—the name and idea are borrowed from Wynne Harlen (2001). A circus involves a set of activity stations set up "around" (hence circus) the classroom. Each station contains a question or questions that students, working in small groups, respond to in the process of performing the activity. The circus provides a fun way to initiate the unit and to find out what knowledge and preconceptions students are bringing to the unit. Instead of a circus, the teacher might choose to use some of these activities as teacher demonstrations to initiate discussion or may choose to have all students perform the same activity followed by discussion. Some of the activities (and everyday examples) might be used periodically for clarifying particularly challenging conceptual situations. Other activities (and everyday examples) might be held back for assessment purposes. A number of these activities can also be used by students later in the unit when they are designing and testing their own investigable questions. A sample circus, including station questions, will be given for each of the five content areas.

The circus section will be followed by a list of sample inquiry questions that students may want to investigate further and in more depth. Only questions that lend themselves to investigation within the limits of classroom supplies are suggested. These so-called investigable questions can be pursued as part of the authentic inquiry part of the unit. Of course, it is hoped that students will be able to generate their own inquiry questions and participate in the full cycle of inquiry, but these sample questions may help to begin that process. Generating questions that can be successfully investigated is indeed one of the most difficult aspects of doing science. It can also be the most rewarding, where students take ownership of their questions, participate fully in the process of doing science, and become scientists themselves.

Acknowledgments

My work with K–12 teachers began many years ago when I was a graduate student and lecturer at the University of California at Santa Barbara. During these formative years, I helped to create and coordinate the UCSB Physics Learning Center, a hands-on science museum that supported university science courses and welcomed K–12 outreach. I thank Doug Scalapino, Bill Walker, Abel Rosales, Rich Harding, and especially Tony Korda for showing me the path.

All of my Middlebury College Physics Department colleagues need to be thanked for supporting my desire to focus much of my professional life on pedagogy and outreach, sometimes at the expense of my science. Also, a very special thanks to Crispin Butler, who, over the years, has given me so many wonderful ideas for engaging physics activities.

Thanks also to my Teacher Education Program colleagues at Middlebury College. Working with you was the best thing I ever did. I have learned so much from all of you, especially on the topics of inquiry-based pedagogy and formative assessment techniques. Keep up the good fight.

To Gregg Humphrey, what can I say? Some twenty-five years ago, Gregg, then Associate Principal of Mary Hogan Elementary School in Middlebury, Vermont, renewed my interest in science outreach. The rest is history. We have been working together ever since—through countless projects, workshops, and institutes on inquiry-based science teaching, learning, and assessment. Gregg joined the Middlebury College Teacher Education faculty a number of years ago and we now coteach the science and math methods for K–6 preservice students. He has always been the "big idea" guy and none of this book would have been possible without his mentoring and support.

Many of the professional development opportunities that I have helped facilitate over the years have been supported by a variety of agencies and institutions. So thanks to the National Science Foundation, the Vermont Department of Education, the Addison Central Supervisory Union, and Middlebury College. I hope the money was well spent.

Thanks to all the wonderful K–8 teachers I have had and continue to have the pleasure of working with. As I said in the dedication, I have greatly improved my college practice by working with you and observing your engaging classrooms. Special thanks to the Addison Central Supervisory Union teachers, and an extra special thanks to Jan Willey, Associate Superintendent of the ACSU, whose enthusiastic support and tireless administrative work made it all possible for so many years.

A number of teachers and colleagues have contributed outstanding leadership to many of our science institutes and workshops: Susan Lewis, Maura Carlson, Graham Clark, and Cathy Byers, among others.

I am indebted to four Middlebury College students: Scott McComb (1993), Ian Smith (1996), Jessica Perkins (1998), and Seth Wolcott-MacCausland (2000). Each of these students worked with me on senior thesis projects that focused on inquiry-based science teaching and learning. Their work has contributed much to this book.

I end by acknowledging what really counts. Thanks to all of the enthusiastic K–8 students who embraced—with delight—the inquiry-based science learning that we have promoted. You have made the road trip so enjoyable and very worthwhile.

PUBLISHER'S ACKNOWLEDGMENTS

The contributions of the following reviewers are gratefully acknowledged:

Elizabeth Hammerman, EdD,
Science Educator and Consultant
Southern Pines, NC

Peggy Rogers, Educator
Burton Elementary School,
Rexburg, ID

Theresa Knaebel, Middle School
Science Teacher
Pittsburgh Public Schools,
Pittsburgh, PA

Nancy H. McDonough, Second
Grade Teacher
Stillman Elementary School,
Tenafly, NJ

Dr. Kerry Williams, Professor
Wayne State College, Wayne, NE

Sheila Smith, Science Specialist
and National Science
Foundation Project Director
Jackson Public Schools,
Jackson, MS

Sally Koczan, Science Teacher
Wydown Middle School, Clayton,
MO

Ken Garwick, Sixth Grade Teacher
Marlatt Elementary School,
Manhattan, KS

Elizabeth F. Day, Middle School
Teacher
Mechanicville Middle School,
Mechanicville, NY

About the Author

 **Bob Prigo** was born and raised in Laguna Beach, California, and attended the University of California at Santa Barbara as both an undergraduate and a graduate student. After receiving his doctorate in Physics and Physics Education from UCSB in 1976, he taught for four years at his alma mater before joining the Physics Department at Middlebury College, Vermont, in 1980. He has been at Middlebury College for the past 27 years. He directed the Teacher Education Program at the College from Fall 2000 to Spring 2006. His research interests include K–16 physics education, holography, physics of sports, and geophysics.

Throughout his career, he has worked and continues to work with K–12 educators on inquiry-based science teaching and learning, supported through grants from the National Science Foundation and the Vermont Department of Education. His work with teachers and schools goes beyond professional development workshops to include phone-call help to local teachers, science assemblies at local schools, lending equipment and resources, and visiting school classrooms to work directly with teachers and students. He was recognized in 1991 as the Vermont Professor of the Year by the Council for Advancement and Support of Education and received Honorable Mention for the Lynton Award for Faculty Professional Service and Outreach in 1999.

From introductory physics courses for liberal arts students to advanced courses for physics majors, and in science methods courses for future teachers, he has always revealed how the study of science is an integral part of a liberal arts education and how the creative mind in science, shared with the arts and humanities, is at the heart of the process of scientific inquiry. He is a strong advocate for an approach to teaching and learning that seeks to open students to the unity of knowledge and finds the creative mind as the central theme that binds the liberal arts together.

Motion and Force

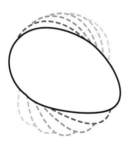

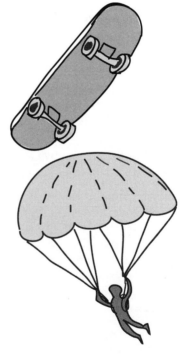

INERTIA AND NEWTON'S FIRST LAW OF MOTION

Concepts

All material objects are stubborn in *two* senses of the word. On the one hand, an object at rest (not moving at all) tends to remain at rest. Examples of this aspect of the concept of inertia are all around us—a book resting on a table, a house resting on the ground. On the other hand, an object moving in a straight line with constant speed (not increasing or decreasing its speed and/or turning) tends to remain moving in the same straight line with the same constant speed. Examples of this aspect of the concept of inertia are a little harder to identify—a spaceship drifting in interstellar space far from gravitating objects, a hockey puck moving freely across the ice. These *two* tendencies make up the *concept of inertia.* Historically, recognizing this duality was a key to unlocking our present understanding of motion. The scientists most responsible for its discovery and clarification, respectively, were Galileo (1564–1642) and Newton (1642–1727).

An object with a larger quantity of matter (say, a jumbo jet) possesses more of these tendencies—that is, possesses more inertia than an object with a lesser quantity of matter (say, a Ping-Pong ball or other table-tennis ball). In other words, both a jumbo jet and a Ping-Pong ball at rest tend to stay a rest, and both a jumbo jet and a Ping-Pong ball moving in a straight line with constant speed tend to remain doing that, but the jumbo jet has much more of these two tendencies than the Ping-Pong ball. The jumbo jet is said to have more *mass, mass being the quantitative (numerical) measure of inertia.*

Mass is also a measure of the amount of "stuff" (matter) that makes up an object. In other words, the mass value for an object gives a numerical value for the quantity of matter as well as for the tendency an object has to remain at rest and for the tendency an object has, if moving in a straight line with constant speed, to remain doing that. In the metric system of units (International System of Units), mass is measured in *kilograms (kg).*

Mass is a distinct concept from and more fundamental than *weight.* Weight is a measure used to quantify the strength of the gravitational force of attraction an object experiences when in the vicinity of the earth, moon, or any other astrophysical body. In other words, weight is a gravitational force and not a measure of inertia. For example, out in interstellar space, far away from any gravitating astrophysical body, a jumbo jet and a Ping-Pong ball would both be *weightless* (would have the same weight of zero) but would still possess greatly different amounts of inertia. In fact, each would have the same amount of inertia (each would have the same mass value) they had on earth or anywhere else, for that matter. Weight (gravitational pull) depends on locality; mass (inertia, stuff) does not. In the metric system (International System of Units), weight is measured in *newtons (N)*. Near the surface of the earth, a 1.0-kg object *weighs* about 9.8 N or about 2.2 lbs.

The concept of inertia sets the *default* for motion. When there is no force on an object, or the net force on an object is zero (the concepts of force and net force will be examined later), the object will be moving in a straight line with constant speed (the speed could be zero and the object not moving). In other word, objects do not need any cause to keep moving in a straight line with constant speed. That is just the way it is. This is the default of nature. As we will see later, forces are responsible for *changing* motion (speeding up, slowing down, and/or turning) but are not needed for straight-line, constant-speed motion. If you could snap your fingers and turn off all the forces in the universe, you would see all the fundamental particles that are the building blocks of matter moving in straight lines with constant speeds.

Indeed, the concept of inertia is the backdrop against which all motion is played out. It is absolutely fundamental to our understanding of motion. Unfortunately, we live in a world dominated by electrical and gravitational forces, so it is natural that children possess the misconception that forces are needed for motion. Indeed, many children hold the misconceptions that constant motion requires a constant force, and that if an object is moving there is a force on it in the direction of its motion, with the amount of motion being proportional to the amount of force (Gunstone & Watts, 1985).

Forces are not required to maintain motion. Straight-line, constant-speed motion just happens on its own without the need for any external agents. Forces (external pushes and pulls) cause a *change* in motion. To be discussed later in more detail, forces cause objects to *accelerate* (speed up, slow down, and/or turn corners).

The concept of inertia is clearly articulated in Newton's First Law of Motion, often called the Law of Inertia: *When there is no force on an object or when the net force on an object is zero, the object will be either at rest or moving in a straight line with constant speed.* The law also works in reverse: If an object is at rest or is moving with constant speed in a straight line, you can conclude either that no forces are being exerted on the object or that the net force on the object is zero.

Before closing on this topic, some interesting extensions of Newton's First Law of Motion must be mentioned, not because they are developmentally appropriate for the student, but only because they should be of philosophical interest to the teacher. The fact that being at rest and being in motion in a straight line with constant speed are equivalent, in the sense that no forces are need for these types of motion, led Einstein to conclude that there is no way to know whether something is actually moving or not. You can feel only acceleration, motion that is changing. In other words, there is nothing you can do and no experiment you could perform to know whether you are absolutely at rest or moving in a straight line with constant speed. Consequently, Einstein argued that the absolute motion of an object through a fixed space is just a false construction of the human mind. There is no such thing as absolute motion and absolute space. This thinking led him to conclude that *all motion must be relative*. You can reference the motion of an object relative to some other object (e.g., the earth moves relative to the sun or a car moves relative to the road), but the absolute motion of an object in some fixed and absolute space does not exist. These ideas, along with some additional reasoning about the speed of electromagnetic radiation, eventually led to the knowledge that both space and time are also relative concepts and to the equivalence of mass and energy, cornerstones of Einstein's Special Theory of Relativity.

Activities

Bowling Ball Versus Nerf Ball: You will need to obtain a bowling ball (more mass and inertia) and a Nerf ball or beach ball (with much less mass and inertia). Place each ball at rest on the ground in front of you. Hit each with a yardstick like you would hit a golf ball. Observe that the bowling ball does not move or moves very little (more tendency to stay at rest) and that the Nerf ball moves quickly away (less tendency to stay at rest). Next, get each ball rolling across the classroom floor and try to stop each with the yardstick. Observe that it is very easy to stop the Nerf ball (less tendency to keep moving) and very difficult to stopped the bowling ball (more tendency to keep moving).

Pendulums: You will need to obtain a variety of balls of various masses (from Styrofoam to steel) that can be hung from individual strings. You will also need to obtain a rubber-tipped dart gun (from your local toy store) and a drinking straw. With the pendulums hanging side by side, shoot each one in turn with the dart gun. Note the motions. The more massive ball will move the least and the least massive one the most. If you want, you can then weigh the balls on a scale to determine their masses to see if you ordered them properly. Instead of using the dart gun, you can "push" on the balls by blowing air at them through a straw. You can also set the pendulums swinging and try to stop them with a dart gun or by blowing air. The more massive pendulum should be harder to bring to rest. You might investigate other ways of providing a controlled push on the pendulums beside the dart gun and blowing air.

Ball on a Truck: You will need to locate a toy truck (or any flatbed cart), a metal ball (or any ball with a reasonable amount of mass), and a small piece of putty. Rest the ball in the center of the toy truck on a small piece of putty. The putty is used to fix the ball to the bed of the truck in order to keep it from initially rolling around inside the truck. It is important that the putty *barely* keeps the ball fixed to the truck. Move the truck forward quickly and observe that the ball wants to stay where it was (relative to the ground) and hits the back of the truck. An object at rest tends to stay a rest. Repeat, but now ease the ball and truck into motion, with the ball moving along with the truck and still stuck to the truck. Stop the truck suddenly by running it into some object (say, a wall). Notice that the ball has a tendency to keep moving in a straight line and hits the forward part of the truck. Repeat, but this time, as the truck and ball are moving together, turn the truck in a sharp circle. Notice that the ball still tends to move in a straight line and hits the side of the truck. You might try this activity using a smooth block (no putty needed) in the truck, instead of with a ball. You might also try scaling up these activities by using a push cart, the kind used for moving supplies around the school, and a larger ball (e.g., a basketball).

Marker Pen Person on a Cart: You will need a marker pen (like a Magic Marker) and a flatbed cart. Balance the marker pen vertically in the center of the cart so it simulates a person in a car. In other words, the maker pen is standing straight up, balanced on its bottom. With your hand, thrust the cart forward and watch the maker pen fall backward. Relative to the ground, the marker pen has a tendency to remain at rest and falls backward. Now repeat, but this time ease the cart into motion—making sure that the pen does not fall over—and run it into some object (like a wall). Observe now that the maker pen falls forward. A marker pen in motion tends to stay in motion. Like the previous activity, this activity is very useful for illustrating the role of seat belts, air bags, padded dashboards, headrests, and child safety seats in automobiles and other vehicles.

Ramp and Slider Investigation: You will need to collect a set of small balls of approximately the same diameter but with different masses. You will also need to construct a ramp (say, from a Hot Wheels track or with a grooved ruler) that runs down to the floor. Position a catcher-slider (anything, like a paper cup, that can catch a ball and slide across the floor will do) on the floor a few inches from the bottom of the ramp. From the same height on the track, release each ball in turn down the ramp and into the catcher-slider. Observe or measure how far the slider moves in each case. The more massive balls have more inertia and tend to keep on moving more than the less massive balls. Other versions of this activity would change the mass of the catcher-slider and/or change the surface over which the catcher-slider slides. Instead of using the slider, you can set up a row of cards (bent to stand up) in front of the ramp and see how many cards each ball can knock over.

Stack and Pull: You will need to cut a two-by-four into seven or more eight-inch-long blocks or locate some wood blocks of similar size. You will also need to sandpaper these blocks smooth. Attach a screw eyelet to one of the blocks in the middle of one of its narrow ends. Stack the blocks directly on top of each other, with the eyelet block at the bottom. Using a string attached to the lower block's eyelet, pull that block quickly out from under the others. Be careful; do not pull the block so hard that it flies into your hand or body or something else in the room. As in the previous activity, the stack of blocks will tend to remain at rest and will not fall over and will remain stacked. You might try different arrangements, like standing some of the blocks vertically or inserting the eyelet block at different places in the stack. Blocks at rest tend to stay at rest.

Stack and Shoot: You will need to locate some circular-shaped playing pieces from a backgammon game. The pieces must be smooth and without ridges (ridged checkers will not work). Stack the pieces (10 or so) on top of each other on a smooth table or floor. Take another piece and, with a flick of a finger, send it across the table toward the bottom piece in the stack. A direct hit will flip out the bottom piece in the stack while the others will fall down in place and remained stacked. The stack of pieces tends to stay at rest. Another version of this activity places a small strip of paper under the stack. The paper can be pulled out quickly without toppling the stack.

Dollar Bill Trick: You will need to locate two plastic water bottles, both filled with water and sealed tightly. You can also use empty soda bottles filled with sand and sealed with a smooth cap. Take a dollar bill and place it between the two bottles balanced on each other, cap-to-cap. In other words, the dollar bill is sandwiched between the two bottles, which are precariously balancing cap-to-cap. One edge, not the middle of the dollar bill, should be between the bottles. Try to pull out the dollar bill quickly without upsetting the balance. One technique that works well is to hold the dollar bill straight out from the bottles with one hand and then quickly karate chop downward with the index finger of the other hand in the middle of the dollar. A bottle at rest tends to stay at rest.

Tablecloth Pull: You will need to locate a smooth piece of cloth (a piece of silk works well) to act as your tablecloth. There should be no seams around the edges. You will also need to locate a set of dinnerware (smooth plate, cup, etc.) and a very smooth table with a sharp edge. Place the tablecloth partway on the table, no more than a foot and a half from the table's edge. Allow the rest of the tablecloth to hang to the floor. Again, make sure the table has a sharp edge and the tablecloth has no edge seam. Set the table. Grab the cloth with both hands in the middle of the cloth hanging to the floor. Pull it quickly *downward.* The table setting should remain at rest.

Jar and Screw Trick: For this barroom trick, you will need to locate a small-mouthed jar and a flathead screw. You also need to construct a plastic ring about six inches in diameter out of a half-inch strip of flexible plastic. Balance the plastic ring on top of the small-mouthed jar with the plane of the ring vertical. Then balance the screw on top of the ring directly above the mouth of the jar. Swipe out the ring with your finger and watch the screw fall into the jar. You must swipe it from the inside of the ring to the outside to make this work. A screw at rest tends to stay at rest.

Inertia Apparatus: You will need to buy this inexpensive inertia apparatus from a science supply house (Inertia Demonstrator, Sargent-Welch). A small piece of cardboard is placed on the apparatus and a ball is placed on top of the cardboard. A flipper mechanism flips the cardboard out from under the ball and the ball remains on the apparatus after the cardboard has been ejected. A ball at rest tends to stay at rest.

Coin and Finger Trick: This is another version of the previous activity, but does not require a commercial apparatus. You will need to cut out a one-and-a-half-inch square piece of thin, smooth cardboard. The thin cardboard from the back of a pad of paper works well. Place the piece of cardboard on the pad of one of your middle fingers. Balance a quarter on top of the cardboard and above your finger. With the middle finger of your other hand, flick out the piece of cardboard without hitting the quarter. The quarter should remain balanced on your finger. A quarter at rest tends to stay at rest. This activity may take a little practice to perfect.

Balloon Puck Fun: You need to buy some inexpensive "balloon pucks" from a science supply house (Puck Set, Carolina Science and Math). These plastic pucks can be lifted off a smooth surface by the balloon that directs air underneath the puck's bottom surface. This provides a cushion of air that greatly reduces the friction between the puck and a smooth table or floor. As long as the balloon is exhausting air, the puck acts like a hovercraft. When launched across a smooth and level table or floor, the puck tends to glide in a straight line with relatively small deceleration. A puck in motion tends to stay in motion.

Toy Hovercraft: You can now buy a relatively inexpensive toy hovercraft from a science supply house (Kick Dis-Air Puck, Educational Innovations). When launched across a smooth level floor, the hovercraft will continue in a straight line at relatively constant speed.

Rotating Table and Pickle Play: You will need to buy a rotating table from a science supply house for these activities (Rotating Platform, Arbor Scientific). A lazy Susan might also work. You will also need a ball (steel ball works best) and a jar of pickles (with pickle juice). Place the ball on the table about halfway out from the center. Rotate the table with your hands back and forth and watch the ball attempt to stay in place. A ball at rest tends to stay at rest. Now attach the ball to the table with a small piece of putty a few inches from the edge of the table. Use just enough putty to keep the ball in place. Ease the table and ball into motion so that the ball is rotating along with the table. Stop the table abruptly and watch what happens to the ball. A ball in motion tends to say in motion and should move off the table in a straight line, tangent to the circle. Place a sealed jar of pickles (with pickle juice) at the center of the rotating table. You will need to remove a few of the pickles from the jar so the pickles in the jar can rotate around freely with the juice. Rotate the table back and forth with your hands. Observe what happens or doesn't happen to the pickles inside the jar. Pickles at rest tend to stay at rest. Now rotate the table steadily in one direction until the pickles and juice start to rotate along with the jar and table. Once the pickles and pickle juice are in motion, reach out and grab the pickle jar with both hands and pick it off the table. Observe the subsequent motion of the pickles. Pickles in motion tend to stay in motion.

Raw Versus Hard-Boiled Eggs: You will need a raw egg (in its shell) and a hard-boiled egg (still in its shell). On a smooth table, spin the hardboiled egg on its side. Stop the egg momentarily by pushing down on it with a couple of fingers and then releasing. The hard-boiled egg will stop spinning. Now repeat with the raw egg. Once released, the raw egg inside the shell will continue to spin some more. The raw egg in the shell in motion tends to stay in motion. See "Rotating Table and Pickle Play," above. This is a good way to tell a raw egg from a hard-boiled one when you are not sure which is which.

C-Shaped Track: You will need to construct a C-shaped wall or barrier on a smooth and flat table. The wall can be made out of a piece of cardboard bent into the shape of a C and fixed to the table with tape. A ball is then made to move against the C-shaped wall, following its curve. The activity is to predict, once the ball leaves the C-shaped path and when the force of the wall is no longer acting on the ball, what path it will follow on the table. Students can predict the path by drawing the anticipated path on a sheet of paper fixed to the table. When the net force on a ball is zero, the ball already in motion will continue in motion in a straight line at constant speed.

Everyday Examples

Parachute Brakes: Dragsters, some jet planes, and the Space Shuttle have parachutes to help them stop. Dragsters and aircraft have lots of inertia (large mass), and once in motion really tend to stay in motion.

Car Inertia: If you are in a car that suddenly accelerates from a stop, your head and body seem to move backward for a moment. An object at rest tends to stay at rest. When you are in a car that is moving along and suddenly decelerates, your head and body seem to move forward. An object in motion tends to stay in motion. When you are in a car that is turning a corner, your head and body seem to move toward the outer side of the car. An object moving in a straight line with constant speed tends to remain doing that.

Whiplash: When a parked car is suddenly hit from behind by another car, a passenger in the parked car can suffer whiplash, where the upper spine is bent backward. A head at rest tends to stay at rest. Headrests in cars are safety features used to avoid whiplash.

Seat Belts, Airbags, and Padded Dashboards: Seat belts, shoulder straps, air bags, and padded dashboards save lives. People in motion tend to stay in motion.

Airplane Flight: When you are inside an airplane that is cruising in a straight line with constant speed, you do not even know you are moving. In fact, there is nothing you can do inside the airplane to know whether or not you are moving. Everything inside the plane is in motion and no forces are needed to maintain the motion of these objects. But when an airplane accelerates during takeoff or decelerates during landing, please fasten your seat belt.

Airport Runways: Runways for large jumbo jets are very long. Jumbo jets are very massive with lots of inertia; thus it is difficult to get them up to takeoff speed and difficult to bring them to rest when landing.

Oil Tankers and Tugboats: An oil tanker full of oil is very massive, so it takes both a long time and distance for it to stop after the engines have been turned off. Once in the harbor, large ships require tugboats to help them stop, turn, and dock.

Magazine Pull: To get the bottom magazine out from under a large stack of magazines, you can apply a quick jerk to the bottom magazine without disturbing the magazines on the top. Magazines at rest tend to stay at rest.

Tightening a Hammerhead: One way to tighten a loose head of a hammer is by striking the base of the handle on a solid surface. A hammerhead in motion tends to stay in motion.

Slip-n-Slide: Sliding on a Slip-n-Slide gives a long ride. With the water film reducing the frictional force, your body in motion tends to stay in motion.

Curling: In the sport of curling, a "stone" is released on a flat surface of ice and tends to move in a straight line with constant speed.

Sand Off a Towel: You can get sand off a towel by snapping the towel. This is done by first getting the sand and towel in motion, and then stopping the towel suddenly. The sand stays in motion and flies off the towel.

Water Off a Dog's Back: A dog shaking water off its back is much like getting sand off a towel. The dog gets the water and its back in motion, and then stops its back and there goes the water.

Water Off a Brush: Shaking a wet paintbrush to remove water is like the dog getting water off its back. Get the water and brush in motion, then stop the brush and there goes the water.

Ketchup Bottle: You can get ketchup out of a ketchup bottle by first moving the bottle and ketchup toward the food and then stopping the bottle. The ketchup keeps moving.

Ice Skating, In-Line Skating, and Skateboarding: Ice skating, in-line skating, and skateboarding are fun because once you get moving, you tend to keep on going.

Icy and Wet Roads: Driving on an icy or wet road can be dangerous because once you get moving, with less friction to stop you, you tend to keep moving in a straight line.

Out of Bounds: As you race to keep the ball from going out of bounds in various sports, it is difficult to stop yourself before going out of bounds. Athletes in motion tend to stay in motion.

Subway Surfing: Standing on a bus, train, or subway ride can be tricky. When starting, stopping, or turning, you must hold on, but when you are moving in a straight line with constant speed, you do not need to hold on.

Skateboarding: Bumping into a curb while skateboarding can be dangerous. Skateboarders tend to stay in motion and fly forward.

Horse Jumping: In a horse-and-rider jumping event, sometimes you see the horse suddenly stop before the hurdle and the rider continue his or her motion over or into it.

Bucket of Water: It is possible to fill a bucket with water (or with something else) and whirl it in a big vertical circle over your head without spilling the water. The bucket continually deflects the water from its tendency to move in a straight line.

Ice Trays: When carrying an ice tray filled with water from the sink to the refrigerator, you often spill water out of the back of the tray when you first begin your movement from the sink. Water at rest tends to stay at rest. When you end your movement at the refrigerator, you often spill water out the front of the tray. Water in motion tends to stay in motion. But you do not often spill the water in the middle of your journey.

Cup Holders and Lids: Liquid in a cup at rest in your car wants to remain at rest; consequently, when you accelerate quickly, the cup of liquid would fall backward if it were not for the cup holder. Liquid in a cup in your car moving in a straight line with constant speed wants to remain doing that; consequently, when a car suddenly decelerates, the cup of liquid would fly forward if it were not for the cup holder. If it were not for the cup holder, when the car turns a corner, the cup of liquid would fly forward and toward the outside edge of the turn, maintaining its straight-line motion.

Cup lids are useful for keeping the liquid from sloshing out of the cup for the same reasons.

Packages in Your Car's Trunk: You place a loose package in the trunk of your car. Similar to the last couple of examples, when you accelerate the car forward, the package hits the back of the trunk. When the car is moving and you hit the brakes and decelerate, the package hits the front of the trunk. When you turn a corner, it hits the side.

Trucks: When riding in the back of a truck (not advisable), you should sit with your back up against the cab. When transporting objects in a truck, you should tie them down to the bed. If you cannot tie them down, then place them up against the cab wall. In this case, they will still tend to move a little when the car accelerates forward or turns a corner. Never allow a dog to ride in the back of a truck.

Bowling: In bowling, it is important to run up before releasing the ball. A ball in motion tends to stay in motion.

Long Jumpers: Long jumpers run quickly before they jump. A jumper in motion tends to stay in motion.

Pizza Oven: To get a pizza into a hot oven, the pizza person moves a large spatula with a pizza on it toward the open oven, then stops spatula, and the pizza keeps moving (a pizza in motion tends to stay in motion). To get a pizza out of a hot oven, the pizza person shoves the spatula quickly under the pizza (a pizza at rest tends to stay at rest) and then draws the pizza and spatula slowly out.

Toilet Paper and Paper Towels: In removing a piece of toilet paper from a fresh roll of toilet paper or a paper towel from a fresh roll of paper towels, a quick jerk tears the paper better than a slow pull. A roll of paper tends to stay at rest, especially when the roll is fresh (more mass).

Earth's Rotation: The earth and other heavenly bodies just keep spinning, no external force is necessary. Rotating objects are said to possess *rotational inertia*.

Cartoons: A common scene in many cartoons shows a character maintaining speed in a straight line, usually going through objects (barns, haystacks, etc.) in the process.

Movies and TV Shows: Many movies and TV shows contain scenes where the concept of inertia is either displayed or misrepresented.

FUNDAMENTAL FORCES AND NEWTON'S THIRD LAW OF MOTION

Concepts

Forces are everywhere. This is especially true of the *gravitational force.* Through the gravitational force, my body is pulled to the earth (and the earth is pulled to my body), the earth is attracted to the sun (and the sun is attracted to the earth), and our sun is attracted to our galaxy of stars (and the galaxy is attracted to our sun). These are manifestations of the fundamental gravitational interaction through which all masses, from particles as small as the electron to objects as large as galaxies, attract each other. The *Universal Law of Gravitation* (Isaac Newton, 1642–1727) posits that a particle with mass attracts any other particle with mass. The strength of the attraction (the force) is the same on both masses and is proportional to the masses of each particle (actually, to the product of the two masses) and grows weaker with increased distance between the particles (actually, as an inverse-square law). Except maybe in the early universe, gravity only pulls, never pushes. Gravity is an attractive force.

The *electromagnetic force* is also everywhere. Through the electromagnetic force, electrons bind to nuclei to form atoms, atoms bind to other atoms to form molecules, and molecules bind together to form solids and liquids. Indeed, this force holds most objects together. But this force can also repel objects. The reason that a chair can hold me up is a result of an overall electrical repulsion between my bottom and the chair seat (the repulsion is mutual, the chair exerts an electrical force on me and I exert an equal electrical force back on the chair). Indeed, essentially all contact forces (objects touching or hitting each other) that we see around us are fundamentally electromagnetic in nature. Friction is an electromagnetic force resulting from the creation and breaking of electrical bonds between two surfaces that have been pushed together (often pushed together by the gravitational force). Electromagnetic forces can be both attractive (pulling) and repulsive (pushing).

The two remaining fundamental forces, the *strong force* and *weak force,* while central to our understanding of processes at the subatomic level, do not manifest themselves directly in the macroscopic world in which we live.

All forces in nature, whether at the fundamental level or when manifested at the macroscopic level, share a beautiful symmetry property: all forces come in pairs. There is no such thing as a single force. What's more, the pairs are of equal strength and always oppositely directed. It is important to note that these force pairs are always exerted on different objects. For example, when I push on a wall with my hand (force exerted *on the wall*), the wall exerts an equal and opposite force back on my hand (force exerted *on my hand*). These force pairs are essential in all types of locomotion—walking, jumping, flying, driving, rockets. See the "Everyday Examples" section below for more details.

Here are three equivalent statements of this symmetry principle, known as *Newton's Third Law of Motion:*

> When two objects "interact" with each other, that is, when two objects exert forces on each other, they always do so with equal strength and in the opposite directions.

> Forces between objects always occur in equal and opposite pairs.

> When object 1 exerts a force on object 2, object 2 exerts a force back on object 1 of the exact same strength but in the opposite direction.

It has become quite common to cite Newton's Third Law in terms of "action and reaction," but this can lead to an unfortunate misconception. One often hears, "For every action there is an equal and opposite reaction." This is misleading because while the forces are in fact the same, the action and reaction may not be the same. For example, if the two objects that are interacting have different masses, then the action and reaction responses will not be the same. The more massive object will respond to the same force with much less change in its motion (less acceleration) while the less massive object will react to the same force with a larger change in its motion (larger acceleration). This important idea will be discussed later under Newton's Second Law of Motion.

Activities

Balloon Bumpers: You will need to purchase or design two bumper carts. You can purchase the bumper carts with spring bumpers from a science supply house (Dynamic Carts, Arbor Scientific). You can make bumper carts by locating two carts (toy cars, skateboards, etc.) with good wheels that allow the carts to move smoothly over a floor or table top. The bumpers can be made from two blown-up balloons (the long type) taped to the front of the carts. Gently crash the carts together and observe the effect each one has on the other. With both carts moving initially toward each other, the equal and opposite forces will stop and reverse the motion of each car. With one cart initially moving toward a cart at rest, the equal and opposite forces will decelerate the moving cart and accelerate the other.

Magnetic Bumpers: You will need two strong magnets and two carts. As in the bumper cart activity, you will need either to purchase the carts or to locate two carts with good wheels. Tape the magnets to the ends of the carts such that the magnets repel one another. Set the carts near each other—with magnets repelling—on a smooth floor or tabletop and release them at the same time. Equal and opposite magnetic forces will cause the carts to fly away from each other. Repeat the activity with the magnets attached to the carts so that the magnets attract one another.

Magnetic Dance: Suspend two magnets vertically from strings (like pendulums) set a few inches apart. Set one of the magnets swinging. Watch the other "dance," and watch how they interact with each other. Magnetic forces are equal and opposite.

Electric Balloons and Tape Repulsion: Hang two small, party-type inflated balloons from thread. Rub them briskly with fur and watch them repel each other electrically. Here the equal and opposite forces are electrostatic. Another version of this uses two 6-inch strips of Scotch tape (or any clear cellophane tape) that have been ripped briskly from a dispenser. Holding the ends, bring the strips close to each other and observe a similar electrostatic repulsion acting on both pieces of Scotch tape.

Remote-Control Car on Movable Road: You will need to borrow a remote-control toy car. You will also need to locate a piece of cardboard (poster board) or a piece of blue board (used for house insulation) and eight or so empty soda cans. The cardboard or blue board should be cut into a piece about 3 feet long and 1 foot wide to make the "road." On a smooth floor or table top, line up the soda cans in a row, on their sides, side by side, and touching each other. Place the road on top of the cans and the car on the road. When ready, remote the car forward and observe the backward motion of the road. The tires push backward on the road (frictional force), and the road pushes forward on the tires. Repeat, but this time remote the car backward.

Turntable Fun: You will need to purchase a rotating turntable (Rotating Platform, Arbor Scientific) for these activities. While standing or sitting on the turntable, throw a massive object (beanbag, bowling ball, etc.) from your side (like you are bowling). Watch your recoil. You push on the object and the object pushes back on you. Also, while standing or sitting on the turntable, hold you arms straight out in front of you and twist the upper part of your body (and arms) one way and watch the lower part of your body automatically twist in the other direction. While on the turntable, pass a heavy object around your body and watch your body rotate in the opposite direction. Place the turntable near a wall. While on the turntable, push on the wall and observe your motion. You push on the wall and the wall pushes on you.

Toy Water Rocket: You will need to purchase a toy water rocket from your local toy store. Follow the directions. The force the rocket exerts in expelling the water is opposite to the force the expelling water exerts on the rocket. You might also try to propel the rocket with air instead of water.

Balloon Sprinkler: You will need to find a small balloon, a plastic soda straw, a needle, and a rubber band for this activity. Use the rubber band to fasten the balloon to one end of the soda straw. Make a 45-degree bend in the straw near the other end. Poke the needle through the center of the straw perpendicular to the plane of the bend in the straw. Blow up the balloon through the straw while holding the needle tip gently between your fingers. Let the air rush out of the balloon and watch the straw spin. A straw exerts a force on the air as the air goes through the turn in the straw, so the air exerts an equal and opposite force back on the straw to make it spin.

Fan Cart: You will need to purchase a toy fan cart from a science supply house (Fan Cart, Sargent-Welch). When the fan is spinning and exerting a force on the air in one direction, the air exerts a force back on the fan in the opposite direction and the cart moves in that direction. Since the spinning blades of the fan cart can be dangerous, this activity should be demonstrated only by the teacher.

Balloon Rockets: You will need to purchase these special rocket balloons from a toy store or from a science supply house (Rocket Balloons, Educational Innovations). Follow the directions on how to inflate the balloon. When you let the balloon go, the air rushes out, making a screeching noise, and the balloon flies around the room. You can also use an ordinary balloon. The long and skinny type of balloon works best. You can also tape an inflated balloon (tied off so the air does not rush out) to a soda straw (straw is taped against the balloon and parallel to its length) and then thread the straw through a wire or string guide that crisscrosses the classroom. When you are ready to launch, cut off the end of the balloon with scissors. The exiting air will propel the balloon quickly along the wire.

Balloon Car Toys: You will need to purchase these toys in a local toy store. They come in various shapes and sizes. A balloon is attached to a toy car. The balloon is inflated and then released. The car is propelled by the air rushing out of the balloon.

Ball and Ramp Recoil: You will need to mount a ramp on a wheeled cart. Watch the cart recoil as the ball rolls down the ramp. As it rolls down the ramp, the ball pushes down and back on the ramp, while the ramp pushes up and forward on the ball.

Two Skateboards and Two Friends: You will need to locate two skateboards and a piece of rope. These activities are done on a smooth floor. With a student standing or sitting on each skateboard, facing each other and holding the rope between them, have each pull on each other and observe the motions. Forces are equal and opposite. Repeat with just one student doing the pulling. Again, the forces are equal and opposite. You can also use just one skateboard with the rope attached to a wall. You can also have the two students push off each other while on the skateboards or, using one skateboard, push off a wall. The forces are always equal and opposite.

Helicopter Balloon: These cheap toys are available commercially (Balloon Helicopter, Arbor Scientific) and can be found in most toy stores. A balloon is inflated and attached to plastic helicopter-type blades. Each blade has a hole running through it with an exit port at the tip and back edge of each blade. When the balloon is released, the air rushes out of the exit ports. This spins the propeller blades (see "Balloon Sprinkler," above). As the blades spin, they push down on the air, so the air pushes upward on the blades and the helicopter takes off.

Jumping Rubber Hemisphere Toy: This toy can be found in most toy stores (Dropper Popper, Arbor Scientific). Turn the rubber hemisphere inside out; it will just stay in this position. Now drop it on the floor, concave side up. When it hits the floor, the hemisphere snaps back to its original, concave-side–down form and jumps off the floor.

Rubber Band Propulsion Car: Mount a rubber band across the flatbed of a cart. Take a small but relatively heavy ball and pull it back on the rubber band like you would on a slingshot. Release the ball and watch the cart recoil. The rubber band pushes on the ball, so the ball pushes back on the rubber band and cart.

Jumping Toys: You will need to go to a local toy store and buy a variety of jumping tops. These windup toys can jump up off a table by pushing down on the table. The table pushes back on the toy and the toy jumps. In some cases, the toy flips as it jumps.

Pushing Bathroom Scales: You will need to locate two identical bathroom scales. You will need a partner. Holding the scales vertically (you should zero the scales in this vertical position), you and your partner push off each other scale-to-scale. When pushing on each other and when all motion (including your arms and hands) between you and your partner has stopped, read the separate scale readings and compare. Repeat by pushing harder. Repeat by pushing softer. The forces, as read off the scales, should be identical in all cases.

Everyday Examples

Stubbing Your Toe: It hurts to stub a toe. When your toe hits the wall hard, the wall hits your toe just as hard.

Gun Recoil: As a gun pushes hard on a bullet, the bullet pushes back hard on the gun.

Boxing: Boxers often hurt their hands. When their hand hits their opponent's face, the face of their opponent hits back.

Firefighters: A hose exerts a force on the water as the water goes through a turn in the hose, so the water exerts an equal and opposite force back on the hose. Multiple firefighters have to brace themselves to hold the hose steady.

Getting Out of a Boat: Getting out of a boat by a dock can be tricky. Your foot pushes back on the boat as the boat pushes forward on you. Since the boat moves, you cannot generate as large a force as you could otherwise.

Running in Sand: Running in sand can be difficult. As your foot pushes back on the sand, the sand pushes forward on your foot. Since the sand moves, you cannot generate as large a force as you could otherwise.

Rocket and Jet Engines: Rocket and jet engines throw high speed exhaust gases out of the rocket or engine with a large force (the thrust), so that a large force will be exerted back on the rocket or engine.

Propellers: Propeller blades push back on the air, so the air will push forward on the blades.

Rope and Rock Climbing: To climb a rope or rock wall, you must pull down to go up. You pull down on the rope or rock in order for the rope or rock to pull up on you.

Walking and Running: To walk or run, your foot pushes back on the ground, so the ground pushes forward on your foot.

Jumping: When you jump, your feet push hard down on the ground, so the ground will push up hard on you.

Bumper Cars: You cannot crash into another bumper car without it crashing into you.

Swimming: When you swim, you push back on the water, so the water pushes forward on you.

Birds Flying: Bird wings push down and back on the air, so the air will push up and forward on the wings.

Handlebars: You pull up hard on the handlebars of your bicycle, so that the handlebars will push down hard on you. This helps you push harder on the pedals and go faster.

Tightrope Walking: You can use your arms to balance on a fence or tightrope. As you begin to fall, you rotate you arms in the direction of the fall and your body will rotate in the opposite direction, so you can regain your balance.

Toddlers: Toddlers often fall backward when they touch an object. When the toddler touches dad, for example, dad will exert a force back on the toddler and the toddler will fall backward. It takes time to get used to Newton's Third Law.

Skiing: To perform a jump turn in skiing, while in the air, you twist the upper body one way and the lower body (plus skis) will turn in the opposite direction. When you hit the snow you will head off in the direction of the skis.

Ballistocardiograph: A heart patient is placed on a special bed that is free to move over a floor (no friction). When the blood is pushed out of the heart, the bed will recoil. The recoil of the bed can be monitored and measured with a ballistocardiograph in order to reveal heart abnormalities.

Kissing: When you kiss softly, you get kissed softly. When you kiss hard, you get kissed hard.

Touching: You can't touch without being touched.

Ice Skating: Push someone on ice skates and you will be pushed in the direction opposite to the direction you pushed.

Trampolines and Diving Boards: When you push down on a diving board or trampoline, it will push up on you.

Jellyfish and Octopi: Octopi and some species of jellyfish can collapse their umbrella-shaped body to push water out in one direction in order to propel themselves in the other.

Moon and Earth: The earth exerts a gravitational force on the moon and the moon exerts the same and opposite force on the earth.

Magnets: When two magnets interact, they push (like poles) or pull (unlike poles) on each other with equal and opposite forces.

NET FORCE, ACCELERATION, AND NEWTON'S SECOND LAW OF MOTION

Concepts

In many cases, more than one force acts on an object. For example, a book resting on a table has two forces acting on it: the downward force of gravity (its weight) and an upward force (contact force) from the table. These two forces cancel each other. While there are two forces acting on the book, the *net force* is zero. When the net force on an object is zero, according to Newton's First Law, the object either will remain at rest (as it does in this case) or it will continue to move in a straight line at constant speed. For another example, consider a jet plane moving through the air with constant speed in a straight line. In this case, there are four forces on the jet. The thrust of the engine produces a forward force on the jet. This force is balanced by a backward force, the frictional air drag on the jet. There are two vertical forces as well. The downward force of gravity (the jet's weight) is balanced by the upward lift force on the wings. When these four forces (thrust, drag, weight, lift) act on the jet, the *net force* is zero. And when the net force is zero, the object either will be at rest or will be moving in a straight line with constant speed (as it does in this case).

Of course, there are many cases for which the net force on an object is not zero. This brings up the question, "What type of motion occurs when the net force on an object is not zero?" To begin to answer this important question, consider a drag-racing car with its wheels spinning and its tires pushing back on the road as it begins its motion down the track. According to Newton's Third Law, the road exerts a forward force on the car (on the tires). This forward force is not balanced by any other force on the race car. While the vertical forces on the race car are balanced (weight downward and road pushing up on the tires), the horizontal forces are not. Indeed, there is a net force in the forward direction on the race car. What is the resulting type of motion? The race car *accelerates*. That is, the race car picks up speed constantly as long as the net force is maintained, going steadily faster and faster as time goes on.

At the end of the race, the race car stops the acceleration by hitting the brakes and releasing a parachute. Again, there is a net force on the drag racer, but now it is in the opposite direction. The friction of the wheels with the ground and the air pushing back on the parachute combine to produce a net backward force on the race car. The net force is opposite to the direction of motion and the race car *decelerates*. In other words, the race car loses speed constantly as long as the same net force is maintained, going steadily slower and slower as time goes on until it comes to a stop.

Generalizing, we can conclude that for motion in a straight line, when the net force on an object is in the direction of motion, the object will accelerate. When the net force on an object is opposite to the direction of motion, the object will decelerate. But what about the case when net force on an object is sideways? What kind of motion occurs in this case? To begin to answer this question, consider a ball rolling off a table. Once in the air, the only force on the ball is gravity pulling steadily down on the object. The upward table force on the ball has been removed once the ball leaves the table. So once off the table, the ball feels only the downward force of gravity. In this case, the force of gravity is the net force. As the ball falls to the floor, the path is a curve. The actual shape of the curved path is a parabola. So when the net force on an object is not in the direction of the motion or opposite to the direction of motion, the object does not follow a straight-line path, but turns. In this case, the object both turns and picks up speed as it falls.

Consider another case, the moon in circular orbit about the earth. The moon experiences a gravitational force of attraction toward the earth. This force on the moon is directed toward the center of the earth. The motion of the moon is perpendicular to the direction of this force. The moon is moving along a circular path and the force (net force) that the moon is experiencing is toward the earth. So, again, when the net force is not in the direction of motion or opposite to the direction of motion, as in this case where it is perpendicular to the motion, the path is a curve. In this case, the path is circular. While the moon does not pick up or lose speed in its circular journey around the earth, the *direction* of its motion is constantly changing. The mathematical definition of acceleration includes not only changing speed (increasing or decreasing) but also changing *direction*. So, in all the cases cited above, from the drag racer to the moon, the object involved is "accelerating." We can conclude that when the net force on an object is not zero, the object accelerates (this includes increasing and decreasing speed as well as turning). The amount of acceleration depends on the strength of the net force, with a larger net force giving a proportionally larger acceleration.

How does the inertia factor (the mass of the object) enter into these situations? Since the mass of an object is a measure of its stubbornness to a change in its motion, it should come as no surprise that a more massive object would experience a smaller acceleration (smaller change in its speed or less turning) than a less massive object under the same net force. Indeed, this is the case. Mathematically, according to Newton's Second Law of Motion, *the acceleration that an object experiences grows proportionally (linearly) with the net force and inversely with the mass.* This can be expressed mathematically as acceleration equals net force divided by mass:

$$acceleration = (net\ force)/(mass)$$

It is interesting to see how beautifully Newton's First Law of Motion is embedded within the Second Law. According to the Second Law, if the net force happens to be zero, the acceleration must be zero. But zero acceleration means that the object does not pick up speed, lose speed, and/or make turns. Consequently, when the net force on an object is zero, the object either will be at rest or will be moving in a straight line with constant speed, Newton's First Law of Motion.

Activities

Note: As you will see, many of the activities described under "Inertia and Newton's First Law of Motion" and "Fundamental Forces and Newton's Third Law of Motion" can be adapted to this section.

Inclined Plane: Take a smooth classroom table and prop up two legs in order to turn it into an inclined plane. Take a smooth ball and let if roll down the table. Observe the acceleration; the ball goes faster and faster as time goes on. In this case, there is a net force on the ball (part of its weight). As the angle of the table gets larger, this net force gets larger. Observe the larger acceleration for larger angles of inclination. Students can also time the ball over a fixed distance to quantify the results.

Galileo's Experiment: You will need to locate balls of various masses that are not affected much by air resistance (small, solid balls work best, but not Nerf, foam, or Ping-Pong balls). Observe the balls as you drop them in pairs from the same height. The balls should accelerate to the ground at the about the same rate and, if released at the same time from the same height, hit the ground at approximately the same time. In this case, the net force on the balls (after being released from the hand) is not zero. In fact, the net force is the gravitational pull (weight). The balls accelerate under this constant net force, picking up speed steadily as they approach the ground. The fact that they hit at the same time is a little subtler. There is a wonderful trade-off that leads to this interesting fact. Consider two balls, one more massive than the other. The more massive ball feels a larger gravitational force of attraction (more weight), and you think this would mean that it should experience a larger acceleration and hit the ground first, but you are forgetting the fact that the more massive ball is also more stubborn (has more inertia) and resists a change in its motion more than the less massive ball. These two factors combine to give the same acceleration for all objects dropped to the surface of the earth, under the condition that air resistance is negligible. This can also be seen directly from the mathematical formulation of Newton's Second Law, acceleration equals net force divided by mass. The net force (weight) in the numerator grows with mass and the inertia factor in the denominator also grows proportionally with mass. This means that the acceleration remains the same, independent of the mass of the object. Students may have seen this phenomenon demonstrated in a science museum where a tube is evacuated of all air and a coin and a feather are dropped together. Both are observed to drop at the same rate.

Book and Toilet Paper: You will need a large book and a piece of toilet paper. Place a rubber band around the book so it will not open. Drop the book from a few feet above the ground and watch it accelerate to the earth. Drop it so it lands on its cover or back. Do the same with a piece of toilet paper. Note the difference. Now place the piece of toilet paper on top of the book and drop them together. With the air resistance now negligible on the piece of toilet paper, both drop at the same rate. See the activity "Galileo's Experiment," above, for an explanation.

Toy Hovercraft: You will need to buy a relatively inexpensive toy hovercraft from a science supply house (Kick Dis-Air Puck, Educational Innovations). You will also need some string and a supply of weights. Tape one end of the string to the hovercraft and place the hovercraft on a level and smooth table. Hang the other end of the string over the edge of the table and hang a weight from the end. With the hovercraft tuned on, release the hover-craft and observe the acceleration. The net force on the hovercraft is the tension in the string produced by the hanging weight. Be sure to catch the hovercraft before it flies off the table. Notice that when the weight hits the ground and there is no longer a net force on the hovercraft, it moves with constant speed in a straight line. Now change the amount of hang-ing weight and note the change in the acceleration. More hanging weight gives a larger net force and a larger acceleration. Now change the mass of the hovercraft (by adding some weight to the hovercraft) and note the smaller acceleration. More mass gives more inertia to the hovercraft and the acceleration is reduced. If you have access to a pulley that you can attach to the edge of the table, place the string over the pulley instead of just hanging it over the table edge.

Wheeled Cart: You will need to locate a flatbed cart with good wheels. You will also need some string and a supply of weights. Tape one end of the string to the front of the cart and place the cart on a level and smooth table. Hang the other end of the string over the edge of the table and hang a weight from the end. Release the cart and observe the acceleration. The net force on the cart is the tension in the string produced by the hanging weight. Be sure to catch the cart before it flies off the table. Notice that when the weight hits the ground and there is no longer a net force on the cart, it moves with constant speed in a straight line. Now change the amount of hanging weight and note the change in the acceleration. More hanging weight gives a larger net force and a larger acceleration. Now change the mass of the cart (by placing some weights on it) and note the smaller acceleration. More mass gives more inertia to the cart and the acceleration is reduced. If you have access to a pulley that you can attach to the edge of the table, place the string over the pulley instead of just hanging it over the table edge.

Fan Cart: You will need to buy a relatively inexpensive fan cart from a science supply house (Fan Cart, Sargent-Welch). You will also need a sup-ply of weights. On a smooth table, with the fan set on the low setting, release the cart and observe the acceleration of the cart. The force of the air pushing back on the blades, according to Newton's Third Law, provides the net force on the cart. Now observe the acceleration with the fan set to the high setting. A larger net force gives a larger acceleration. Now add some weight to the cart in order to increase its mass. Observe the slower acceleration. The spinning blades of the fan cart can be dangerous, so this activity should be demonstrated only by the teacher.

Magnetic Cart: You will need to locate a flatbed cart, two strong magnets, and a set of weights. Tape one magnet to one end of the cart. Use the other magnet, held in your hand, to repel (push) the cart across a smooth table. Try to keep the two magnets at about the same distance from each other in order to maintain a relatively constant net force on the cart. Observe the acceleration. Repeat, but now try to hold the two magnets closer to each other to establish a larger net force. Observe the larger acceleration due to the larger magnetic force. Now add some weights to the cart to increase its mass. Repeat and note the smaller acceleration for the more massive cart.

Everyday Examples

Car Versus Truck: A small car (less mass) and a massive truck are side by side waiting for the light to turn green. The car's acceleration off the start will be much larger than the truck's acceleration because of the car's much smaller mass.

Bowling Ball Versus Baseball: You can accelerate a baseball in your hand up to large speeds. A professional baseball pitcher can accelerate a baseball from 0 mph to 90 mph before releasing it. Not so with a bowling ball. Because of its much larger mass, a bowling ball will undergo a much smaller acceleration and cannot be released nearly as quickly.

Drag Racer: A drag racer accelerates down the track due to the net force on it established by the forward frictional force on the tires.

Car Skidding to a Stop: A car slams on its brakes. The tires lock, and the frictional force from the road is exerted back on the tires and decelerates the car. You often see tread marks left on the road.

A Ball on a String: You whirl a ball on the end of a string. The tension in the cord provides an inward force on the ball and the ball moves in a circle (accelerates). Furthermore, if you let the string go, the ball flies off tangent to the circle, revealing its inertial tendency to travel in a straight line.

Karate: In breaking boards, a karate master's hand decelerates into the wood in a very short time. This gives a very large deceleration, so the force the wood exerts on the hand is very large. Consequently, by Newton's Third Law, the force the hand exerts on the wood is also very large, large enough to break a stack of wood.

Jet at Takeoff and Landing: A very large force (from the jet engines) is required to accelerate a massive jet to takeoff speeds. The same is true for landing. In landing, the jet engines are reversed and the brakes applied in order to decelerate the jet from its landing speed to zero in the distance provided by the runway.

Space Shuttle Liftoff: During liftoff, the thrust from the solid rocket boosters must provide a force on the spacecraft (space shuttle plus boosters) that is greater than the weight of the spacecraft. The net force that accelerates the spacecraft is the thrust force minus the weight.

Landing on the Moon: In order for a spacecraft to land on the moon, it must fire its thrusters downward to make the force on the spacecraft upward (Newton's Third Law). This upward force must be larger than the downward weight (which is about one-sixth the value it would be near the surface of the earth) in order to establish a net upward force on the spacecraft. This net upward force, opposite to the downward direction of motion, will decelerate the spacecraft.

Jumping: In order to jump off the ground, your legs must exert a force down on the ground that is larger than your weight. In this case, the force the ground exerts upward on you (which is the same strength as the force your legs exerted on the ground, Newton's Third Law) will be larger than your downward weight. This gives a net upward force that will accelerate you off the ground. Once you leave contact with the ground, the net force is only your weight (downward) and you will decelerate until you reach the top of your jump. On the way down, your weight is in the same direction as the motion and you will accelerate back to the ground. Can you jump higher on the moon? Why?

Bat and Ball: The contact force that the bat (tennis racket, etc.) exerts on the ball first decelerates and then accelerates the ball.

Planets: Objects have a tendency to move in a straight line with constant speed. For the planets going around the sun, the gravitational attraction to the sun gives a net inward force on the planet (toward the sun) that continually deflects (accelerates) the planet into its circular path (elliptical, to be exact) around the sun.

Raindrops and Skydivers: When a raindrop begins its fall to the earth, the only force (the net force) on it is its weight (downward). It accelerates, but as it picks up speed, the air resistance force on it grows in the opposite direction (upward). At some point in the motion, it has gained enough speed (called the "terminal speed") to make the air resistance force equal to the weight, making the net force zero. At this point the raindrop moves with constant speed. This happens at relatively slow speeds for raindrops (this is good, or raindrops would hurt when they hit). The same happens for skydivers, but the terminal speed is much higher. Skydivers can spread their bodies to increase the air resistance force to decrease the terminal speed. Of course, a parachute greatly increases the air resistance to give a safe terminal speed.

Woodpeckers: When a woodpecker's beak, moving at high speed, hits a tree, the tree exerts a force on the beak. This force decelerates the beak and head of the woodpecker. Since the loss of speed happens over a very short period of time, this deceleration is very large and, consequently, so is the force on the woodpecker. The woodpecker shuts its eyes to help keep them from flying out of its head (think object in motion wants to stay in motion). Also, its brain wants to stay in motion. The brain cavity has evolved to protect the brain from this rapid deceleration (think safety helmet).

Amusement Park Rides: Amusement park rides are designed for accelerations—quick accelerations, decelerations, and turns. In this way, your body feels a variety of forces, including gravity, that give you the thrills.

Springs: Springs can be used to accelerate and decelerate objects. When an object hits a spring, the spring compresses and exerts a force back on the object that causes the object to decelerate—think shock absorber. When an object is compressed onto a spring and then released, the spring exerts a force in the direction of motion and the object accelerates—think pinball release. A pogo stick spring does a little of both, decelerating a person on the way down and accelerating the person on the way back up.

Turning a Corner: When a car or some other wheeled vehicle turns a corner, the net force is the frictional force that the road exerts on the tires. This force is sideways on the tires and directed toward the inside of the curve, that is, toward the center of curvature of the corner. If the road is icy, the road is wet, there is loose gravel on the road, or the tires are too smooth, this frictional force will be greatly reduced. In this case, the car will have trouble negotiating the turn and will continue in a straight line and head off the road. When a person is running and turning a corner, a similar situation occurs. The net force on the feet is friction, and it is directed toward the center of curvature. It is difficult to turn a corner on a smooth floor with socks on your feet.

MOTION CIRCUS

The following set of ten activities, selected from the activities described in the last three sections, could be used to begin a unit on motion. These activities would be set up around the classroom in a circus format. Next to each activity, a simple description of how each activity is to be performed would be displayed, along with a question or questions to be answered by the student in conjunction with performing the activity. Obviously, the teacher will need to rewrite these descriptions and questions to make the language and analysis appropriate for the grade level. It is suggested that students work in pairs or small groups. One option would be to have students perform the activities a few at a time and run the circus over a few days. Another option would be to use some of these activities as teacher demonstrations for whole-class discussion. In any case, students should be encouraged to probe the activities beyond the descriptions and initial questions and begin to think of additional questions they might want to investigate on their own later in the unit.

1. Book and Toilet Paper

Drop the book from a few feet above the ground (drop it so that it hits on one of its flat sides). Drop one sheet of toilet paper from a few feet above the ground. Now place the sheet of toilet paper directly on top of the book and drop the two together. What do you think is the cause of this difference?

2. Marker Pen Person on a Cart

Place the marker pen at the center of the cart so that it is standing straight up, cap up, with the cart at rest (not moving). Pull or push the cart forward or backward rapidly as if you were trying to "pull the rug out from under the marker." Observe what happens to the marker.

Now place the marker pen at the center of the cart so that it is standing straight up, cap up, and carefully ease the cart into motion so that the marker pen remains standing as it moves along with the cart. Stop the cart by having it "crash" into something (like your hand) and observe what happens to the marker pen. What do you think is the difference in the two cases?

3. Jumping Toys

Find a way to make each of these toys jump up from a smooth floor or tabletop. When you jump off the ground, in what direction do your feet push on the ground? Is the same sort of thing happening with these toys?

4. C-Shaped Track

Predict the shape of the path the steel ball will take when it leaves the C-shaped curve. Now roll the ball against the inner surface of the C-shaped curve and see what path it actually takes when it comes off. What was the shape of the path? How did your prediction compare to reality?

5. Pushing Bathroom Scales

You and your partner will use the bathroom scales to push on each other. When pushing on each other and when all motion (including your arms and hands) between you and your partner has stopped, read your separate scale readings and compare. Repeat by pushing harder. Repeat by pushing softer. What did you find out?

6. Turntable Fun

A. Carefully stand or sit on the turntable. With the turntable not moving, hold out your arms in front of you and twist the upper part of your body (including your arms) one way and then the other. How did the lower part of your body and the turntable react to this motion?

B. Carefully stand or sit on the turntable. With the turntable not moving, take the heavy object and carefully but swiftly pass it from one hand to another *around* your body in one direction. Try the other direction. If the object is too heavy for you, just take note of somebody else's result. How did your body and the turntable react?

C. Carefully stand or sit on the turntable. With the turntable not moving, take the heavy object and throw it. Throw it hard and outward (using one hand) from one side of your body like you were bowling a bowling ball. Try throwing it with your other hand from the other side of your body. How did your body and the turntable react?

Identify a common pattern in the "reactions" in all three of these exercises.

7. Ramp and Slider Investigation

Weigh each of the five balls. Release one of the balls (always from the same height on the track) and measure how far it moves the slider across the floor. Repeat for the other four balls. Make a chart showing how far the sliders moved compared to the weights of the balls. Do you notice a pattern?

8. Remote-Control Car on a Movable Road

Place the cardboard road on top of the soda can wheels. Place the remote-control car in the middle of the road. Predict what will happen when you remote the car forward. Predict what will happen when you remote the car backward. Now try both and see what happens. What do you think is going on here?

9. Stack and Shoot, and Dollar Bill Trick

A. Stack the smooth backgammon pieces on top of one another on a smooth table or floor. Take one of the pieces and slide it or flick it rapidly toward the bottom backgammon piece in the stack. See if you can remove this piece without having the stack collapse.

B. Place one edge of the dollar bill between the two bottles balanced on one another, cap-to-cap. Hold the dollar bill straight out with one hand and, with the pointing finger of your other hand, karate chop downward on the dollar bill to see if you can remove it without upsetting the balance of the bottles.

What is common to these two events?

10. Toy Hovercraft

Turn on the toy hovercraft. Place it on a smooth table or floor and give it a small push and let it go. What type of path does it follow? What keeps it going?

SAMPLE INVESTIGABLE QUESTIONS

• *Ramp and Slider Investigation:* How would the distance that the slider moves change when the height of release is changed? How would the slider distance change when the sliding surface is changed (carpet, cloth, etc.)?

• *Remote-Control Car on Movable Road:* What would happen if the car were on the movable road and the car turned a corner (instead of going straight forward or backward)? Which way would the road move in this case? The road should be sitting on some marbles instead of the soda cans.

• *Toy Hovercraft:* On what types of surfaces will the hovercraft work? On what types of surfaces will it not work? How much weight can the hovercraft hold and still fly?

- *Jumping Toys:* How well do these toys jump off different surfaces? How does adding some weight to the toys affect how high they jump?

- *Fan Cart and Hovercraft:* Can the fan cart pull the hovercraft (or wheeled cart) across a floor? How does pulling an object affect the acceleration of the fan cart? How well can the fan cart pull objects over various types of surfaces? Since the spinning blades of the fan cart can be dangerous, this activity should be demonstrated only by the teacher.

- *Bouncing Balls:* What kinds of balls bounce the best?

- *Pendulums:* How does the length of a pendulum affect the "period of a pendulum" (how long it takes to complete one full cycle of the motion)? How does the mass of the pendulum bob affect the period? How does the height from which the pendulum bob is released affect the period? How long a pendulum do you need to have a one-second period?

- *Balloon Rockets:* What type of balloons work best as a balloon rocket? How does the amount of air in the balloon affect its acceleration? Can you make a balloon rocket where the balloon flies straight up?

Fluids and Buoyancy

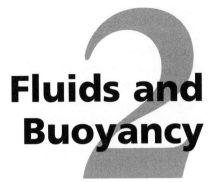

GASES AND LIQUIDS AS FLUIDS

In the conceptual discussion that follows, the word *molecule* will be used to refer to *both* an atom (basic unit of an element in nature) and a molecule (two or more atoms bound together to form a composite entity).

Gases like air and liquids like water are considered separate states of matter, with different physical properties. Gases are generally low in density with their molecules far apart from each other, while liquids are much more dense, with their molecules essentially touching each other. A gas tends to fill up the space it occupies (air in a room), while liquid in a container (water in a glass) fills a space only up to a specific level (with gravity present). Gases are relatively easy to compress, while liquids are very difficult to compress.

But gases and liquids also have some very important properties in common. Both gases and liquids flow—winds and rivers. Flowing gases and liquids can exert forces—wind power and waterpower. Objects can "fly" in both media—birds in air and fish in water. Both gases and liquids exert pressure on objects placed within them or on the walls of the container they occupy—air pressure and water pressure. Both gases and liquids provide buoyant forces on objects placed within them—hot air balloon floating in air and boat floating in water. Because gases and liquids share many important properties and because many of these properties can be understood in terms of the same physical principles, it is both convenient and common practice to place both gases and liquids within the single category of *fluids*.

PRESSURE IN A GAS AND ATMOSPHERIC PRESSURE

Concepts

The large numbers of molecules that constitute a gas (like the air in a room, primarily nitrogen 78%, oxygen 21%) are in constant motion. Sometimes these molecules bounce off (interact with) each other, but more often than not they miss each and reflect off objects placed within the container and off the walls of the container. The molecules are not all moving with the same speed or in the same direction. In fact, the motion is extremely chaotic, with molecules moving with a spectrum of speeds in quite random directions. The *temperature* of the gas gives a statistical measure of the speed of the molecules. In fact, the temperature of a gas gives a measure of the *average speed* of the molecules. Higher temperatures mean higher average molecular speeds. For this reason, the molecular motion of a gas is often referred to as the *thermal motion* of the gas.

At room temperature in air, the average speed of the molecules is quite large—on the order of hundreds of miles per hour. Under normal conditions, the number of molecules is also quite large—approximately 25 million billion per cubic centimeter. Since the number of molecules is so incredibly large and their motion so completely chaotic, in the process of hitting and bouncing off the surface of an object and off the walls of a container, the molecules exert a *very steady* pressure on the object and container walls. In fact, under normal conditions for air in a room, there are approximately 300,000,000,000,000,000,000,000 individual molecular collisions on each square centimeter of surface area every second. This gives rise to a steady pressure on any object placed within the gas and on the sides of the container occupied by the gas.

To be more specific, the *force per unit area* exerted by the gas on an object and container walls is called the *pressure of the gas.* In the metric system (International System of Units), pressure is measured in units of "newtons per square meter" (N/m^2), called a pascal (Pa). In the United States, pressure in most often measured in units of "pounds per square inch" (psi), where $1 \text{ psi} = 6.9 \times 10^3 \text{ Pa}$. It should be mentioned that a common practice is to discuss the "pressure of a gas" even without reference to an object placed within the gas or to the walls of the container. But to measure the pressure, an object must be immersed in the gas (a pressure gauge, for example).

If the gas is atmospheric air, the gas pressure is often called the *atmospheric pressure* or *barometric pressure.* Under normal conditions, the atmospheric pressure at the surface of the earth is a little less than 15 pounds per square inch (15 psi). This pressure varies a little from day to day, depending on the comings and goings of weather fronts. It also varies with elevation, with the pressure decreasing as one ascends through the thinning and colder atmosphere. It is important to realize that 15 psi is a very large pressure. To see this, consider a 10"-by-10" piece of paper sitting on your desk. Bathed in the atmosphere, this piece of paper experiences about 1500 pounds of force on each of its sides! In this case, as in most cases, the atmospheric pressure is the same on *all* sides of an object, so the object feels *no net force* due to this pressure. In other words, in most cases, we do not even notice this large pressure. Only when the equilibrium is changed and one part of an object experiences a different pressure from some other part will a net force result. If no other forces are balancing this pressure difference, the object will move (accelerate).

In order to discover the factors that determine the magnitude of the pressure of a gas, consider a container of gas, say, a can filled with air. Surely the *number of molecules* in the gas is important. It makes sense that if more molecules were added to the can that the pressure would increase, or if some molecules were removed from the can that the pressure would decrease. A molecular argument would reason that if more molecules were added to the can, without any changes in the molecular speeds (temperature) and the size of the can (volume), there would be *more collisions per second* with the walls of the can.

It is just as obvious that the magnitude of the pressure depends on the *temperature* of the gas in the can. Since temperature gives a measure of the average speed of the gas molecules, higher temperature means that the molecules (on average) are moving faster. This means that the molecules would hit the walls of the container *harder* and *more often*, assuming that the number of molecules and the size of the can are not changed. Of course, a decrease in temperature would mean the molecules are moving slower, not hitting as hard or as often, and this would lead to a decrease in the pressure of the gas.

The third and last factor that is important in determining the pressure of a gas in the can is not so obvious: the *volume* of the can (the volume of the gas in the can). But here the dependency is an inverse relationship. If the volume of the can were increased, without changing the number of molecules and temperature, the pressure of the gas would decrease. This would happen because with the larger volume, the molecules must travel a greater distances within the can before they hit the walls. This means they will hit *less often* and that the pressure would be reduced. Furthermore, if the volume increased, then there would be fewer molecules hitting per unit area, since the surface area would increase with the volume increase. This would contribute to a decrease in pressure. If the volume of the can were decreased, the argument would be that the molecules have less distance to travel and would hit *more often* and the number hitting per unit area would be *more*—both factors contributing to a higher gas pressure.

In summary, the pressure of a gas (symbol P) depends on only three factors associated with the gas: the number of molecules (symbol N) that make up the gas, the temperature (symbol T) of the gas, and the volume (symbol V) of the gas. We can say that P grows directly with N and T, and inversely with V. In algebraic form, we can express these relationships in a single mathematical expression (often called the Ideal Gas Law):

$$P = kNT/V$$

The parameter k is a mathematical constant. This quantitative relationship works quite well for gases that are not too dense or complex. The scientists most responsible for its development were Robert Boyle (1627–1691), Jacques Charles (1746–1823), and Amedeo Avogadro (1776–1856).

Activities

Suction Cup Fun: You will need to locate suction cups of various sizes. You should also purchase two large suction cups of the type used to pull out car door dents or to carry panes of glass (Atmospheric Pressure Cups, Arbor Scientific). You might also want to purchase a couple of plumber's helpers (toilet plungers). Stick two suction cups together. By forcing the air out of the space between the cups, you reduced the number of molecules (N). When the suction cups spring back, the volume (V) is also increased. Both of these factors (smaller N and larger V) reduce the pressure inside of the cups. Since the atmospheric pressure outside the cups has not changed, the pressure difference is enough to hold the cups firmly together. You can try to "stick" the suction cups to various surfaces (a window, a table, your forehead, etc.). Try repeating the activity after making a small hole in one of the suction cups.

Rubber and Soda Can Mats: You will need to buy one of these amazing atmospheric pressure devices from a science supply house. One of the devices is just a square piece of rubber with a handle in the center (Atmospheric Pressure Mat, Arbor Scientific). When you place this rubber mat on a smooth surface and attempt to pull it up off the surface using the handle, you can't. As you begin to pull on the mat, the volume of air trapped under the mat increases (V increases). This reduces the pressure under the mat compared to the atmospheric pressure on top of the mat. In fact, the pressure difference will increase the more you pull up on the handle and increase the volume. The same idea has now been used to hold a soda can on a table (Lil'Suctioner, Educational Innovations). You place a soda can into the circular mat, and the pressure difference will "hold" the soda can to the table.

Crushing a Bottle: Locate an empty plastic milk or soft drink bottle. Stick your mouth over the opening and suck out the air. This can be accomplished by first sucking the air out through the mouth, followed by exhausting the air through the nose. If you repeat this procedure without taking your mouth off the bottle, you can eventually crush the bottle. By reducing the number of molecules (N) inside the bottle, the pressure inside the bottle is lowered. The higher atmospheric pressure outside the bottle will crush the bottle.

Inexpensive Vacuum Canister: You will need to purchase special canisters that have been designed to prevent food spoilage (Student Vacuum Pumper and Chamber, Arbor Scientific). These relatively inexpensive canisters come with a lid (with a tight seal) and a small handheld pump for removing the air from the canister. Try placing various things inside the canister (partly inflated balloon, marshmallows, sealed bag of chips, sealed clear zip-type food storage bag, bubble wrap, whipped cream, shaving cream, etc.) and removing the air. Removing the air from the canister reduces the number of air molecules (N); as a result, the pressure inside the canister is reduced. Since the pressure inside the bag of chips, for example, has not been changed, the bag of chips will expand. If you continue to pump on the canister, you can expand the bag of chips to the point where the bag pops open. Wait to see what happens with marshmallows, whipped cream, and other objects you place in the vacuum canisters. Have fun.

Syringe Play: You will need to locate or purchase some plastic syringes (Disposable Syringes, Carolina) and a piece of rubber or plastic tubing (Tubing, Carolina). Obtain syringes of various sizes. The tubing needs to fit snuggly over the syringe openings. Take a syringe and hold a finger over the opening. If the syringe comes with a cap that seals the opening, you can use this cap instead of your finger. Move the plunger *in* with your thumb and feel the increase in pressure as you decrease the volume (decrease V) of air trapped in the syringe. Let the plunger go and it will move back to its original position. Now pull the plunger *out* (increase V) and feel the pressure difference that wants to pull the plunger back to its original position. Now connect the openings of two syringes with the piece of rubber tubing. Make sure the rubber tubing fits tightly over the openings. With a syringe in both hands, push in on both plungers and feel the pressure increase (due to decreasing V). Release. Now push *in* on only one plunger and watch the other plunger move *out*. Pull *out* one of the plungers and watch the other plunger move *in*. This is the basic principle of hydraulic systems (except a liquid is used in place of the air) that can be used to move objects.

Turkey Baster: You will need to locate a turkey baster. Squeeze the rubber bulb on the end of the turkey baster and then insert the tip (the open end) in a cup of liquid. Release the bulb. Liquid will rise up into the baster. When the bulb is released, the volume of the air trapped inside of the baster increases (V increases). This lowers the pressure inside the baster. The atmospheric pressure pushing down on the surface of the liquid can now push the liquid into the baster. Now, to get the liquid out of the baster, squeeze the bulb and decrease the volume (V decreases). This increases the pressure and pushes the liquid out of the baster. Of course, the same principle applies to eyedroppers and squirt guns, as well as to a number of other devices that are used to pick up and squirt liquids.

Popguns: You will need to purchase a popgun from a toy store. Most toy pop-guns are constructed with a cylindrical air cavity and plunger. The ball (or some other projectile) is inserted into the end of the gun. When the plunger is pushed, the volume of the air cavity is reduced (V decreases). This increases the pressure inside the gun and the ball shoots out.

Playing With Straws: You will need to locate some drinking straws and a cup of liquid. Pick up a straw and place it vertically into a cup of liquid. Lift the straw and the liquid will run out of it. Now insert the straw back into the cup of liquid, but this time place a finger over the hole in the straw before you lift the straw out of the liquid. A drop or two of liquid will run out of the straw, but most of the liquid will not fall out. With the straw sealed with your finger, the volume of the air trapped inside of the straw will begin to increase (V increases) as some of the liquid begins to fall out. This reduces the pressure inside the straw. At some point, the air pressure inside the straw will be reduced enough so that the atmospheric pressure outside the straw can hold up the liquid in the straw. If you now take your finger off the hole, the liquid will fall out again. Try drinking through a straw when a second straw, open to the atmosphere, is also inserted in your mouth. Now it will be more difficult to reduce the pressure in your mouth cavity and, consequently, more difficult to draw liquid into your mouth.

Old Faithful: You will need to locate a soda straw, a small soda bottle, and some clay. Put the soda straw into the soda pop bottle that's half-filled with water. Make sure the bottom of the straw is below the water's surface. Seal the bottle opening around the straw with clay. Now blow through the straw into the bottle. If the seal is tight, you should be able to blow only few bubbles' worth of air into the bottle. After a few bubbles of air have been blown into the bottle, take your mouth off the straw and step back. Some water should come shooting out of the bottle. Blowing some air into the bottle increased the number of molecules of air inside the bottle (N increased). This increased the air pressure inside the bottle. This increase in air pressure pushes some of the water up and out of the straw. Instead of a straw, a soda bottle, and a piece of clay, you can substitute a glass tube, a glass beaker, and a rubber stopper with a hole to do the same activity. The glass tube must fit snuggly through the hole of the stopper, and the stopper must fit snuggly into the mouth of the beaker.

Funnel Fun: You will need to locate a funnel, a soda bottle, and some clay. Place the narrow end of a funnel into a mouth of the bottle. Seal the opening of the bottle around the funnel with the clay (this seal must be airtight). Pour water into the funnel and observe. Some water will go through the funnel, but soon the water will back up in the funnel and not go through the opening. This happens because as some water goes through the funnel and enters the bottle, the volume of the air inside of the bottle is decreased (since the water is taking up some of the space). This decrease in volume (V decreases) leads to a greater air pressure inside the bottle. This increase in air pressure is large enough to keep more water from entering the bottle.

Sandwich Bag in a Cup: You will need to locate a plastic sandwich bag, a clear plastic drinking cup, and a rubber band. Insert the sandwich bag inside the cup—like you would a trash can liner inside a trash can—with the edge of the sandwich bag hanging over the edge of the cup. Use a rubber band to seal the sandwich bag edge (the part of the sandwich bag that is hanging over the edge of the cup) to the outside of the cup. This seal needs to be airtight. If a rubber band does not provide an adequate seal, try duct or masking tape. Now reach into the cup and try to pull out the baggie with your fingers. If the seal is tight, the baggie will resist being pulled out of the cup. As you pull on the baggie, you increase the volume (V increases) of air between the baggie and the cup. This reduces the pressure in the air between the sandwich bag and the cup compared to the atmospheric pressure outside the sandwich bag. The more you pull, the larger the volume, the lower the pressure, and the harder it is to pull out the sandwich bag. This activity can be scaled up to the size of a trash can. Place a trash can liner inside a trash can and seal it to the outside of the trash can. Again, you will need a good seal—try duct tape in this situation.

There's a Hole in My Bucket: You will need an empty plastic soda bottle with a cap. Punch a small round hole near the bottom of the bottle. Make the hole big enough so that when the bottle is filled with water, a steady stream of water exits the bottle. While holding your finger over the hole, fill the bottle with water. Then screw the cap tightly onto the bottle. Release your finger and observe. Some water will exit the hole, but soon the water stream will stop flowing altogether. As some water flows out through the hole, the volume of the air cavity inside the bottle increases (V increases). This reduces the pressure inside the bottle. The atmospheric pressure outside the bottle is large enough (once the pressure has been reduced inside the bottle) to stop the water from exiting. If you punch a second hole near the top of the bottle, even with the top on, the water stream will, once again, flow out of the bottle continuously. With the second hole, the pressure in the air inside the bottle cannot be reduced. The pressure cannot be reduced because air will enter through the top hole to maintain atmospheric pressure inside the bottle. This is why containers that are designed for liquid to be poured or drunk have a second hole or opening—to allow the liquid to flow out easily.

The Upside-Down Glass: You will need to locate a drinking glass and a piece of thin cardboard or tagboard. Fill the glass of water to overflowing. Place a square piece of cardboard (cut to be a little larger than the mouth of the glass) on top of the glass. While holding the cardboard firmly against the rim of the glass, turn the cup upside down. It is best to do this over a bucket or sink. Now slowly let go of the paper. With a little practice, the cardboard should stay in place. After you release the cardboard, the water drops a little and begins to push against the cardboard. You might notice an outward budge in the cardboard. As the water drops a little and bulges out the cardboard, a small space is established between the water and cup. This space is a volume with only a few air molecules (small N) and increasing volume (larger V). This means that the air pressure in this space is very small and the atmospheric pressure pushing up on the cardboard is large enough to support the water in the glass.

The Expanding and Collapsing Balloon: You will need to locate an empty soda bottle and a balloon. Stretch the mouth of the balloon over the mouth of the empty soda bottle. Place the bottle in hot water (or wrap it with a hot towel or with a heating pad) and watch the balloon expand. An increase in the temperature of the air in the bottle (T increases) increases the pressure of the air inside the bottle—and the balloon expands. Now place the bottle in a bath of ice water and observe the balloon as it collapses into the bottle. A decrease in the temperature of the air in the bottle (T decreases) decreases the pressure of the air—and the balloon is pushed inside the bottle.

Bubbles and Rising Water: You will need a soda bottle, a clear soda straw, a bowl, food coloring, a piece of clay, and a cloth wrap. Place the straw half inside and half outside the mouth of the bottle and seal the opening around the straw with clay. This needs to be an airtight seal. Fill the bowl with water (at least 3 inches deep) that has been darkened with the food coloring. Hold the bottle above the dish with the straw submerged in the colored water. Wrap a hot towel around the bottle. As the air warms in the bottle, you should observe bubbles in the water coming from the straw. As the temperature increases in the bottle (T increases), so does the air pressure, enough to force some air out of the bottle and into the water (bubbles). Now try the reverse. Wrap the bottle with a cold cloth and observe that some water rises in the straw. As the temperature of the air inside the bottle decreases (T decreases), so does the pressure. The atmospheric pressure on the surface of the water is now larger than the inside pressure and can raise the water in the straw.

Burping Dime: You will need to locate a glass bottle with a mouth opening on which a dime can sit without falling through. You will also need a hot plate. Put a small amount of water in the bottle. Place a dime over the mouth and rub some water around the edge of the dime to create a water seal. Heat the bottle on the hot plate. If the seal is good, the dime will "burp" repeatedly. As the temperature rises in the bottle (T increases), so does the air pressure in the bottle. With a good seal, the pressure will become large enough to lift the dime.

Everyday Examples

Tires, Balloons, and Basketballs: When you pressurize objects such as tires, inner tubes, air mattresses, balloons, basketballs, beach balls, and tennis shoes, you either blow or pump more air into them. By adding more molecules (N increases), the pressure increases.

Peashooters and Blowguns: When you blow into one end of a peashooter or blowgun, you increase the number of molecules in the "barrel." This increases the pressure and pushes the pea or dart out of the barrel.

Collapsing Cheeks: When you suck air out of your mouth cavity and down into your lungs (keeping you mouth and nose shut), notice that your cheeks collapse. With the reduced pressure in your mouth, the atmospheric pressure on your cheeks pushes them in. You can also do the opposite by keeping your mouth shut and pushing air from your lungs into the mouth cavity. Now your cheeks will bulge out.

Breathing: To bring air into your lungs, a diaphragm expands a cavity in your chest and increases the volume of your lungs. This volume increase decreases the pressure so that air from the atmosphere, at higher pressure, will enter the lungs. When the diaphragm contracts the cavity, the lung pressure increases, pushing air out of the lungs.

Hand Noises: Place your hands together to form a pocket of air between your palms. Squeeze this pocket of air (V decreases) and the pressure will increase, enough so that air will rush out of your hands and make a "funny" sound. You can also do this with other parts of your body, for example, by cupping a hand under an armpit and then bringing your arm down, or by cupping a hand over an eye socket and squeezing the air out.

Popping a Bag: You can pop a bag by first blowing air into it, then sealing off the air, and finally hitting the bag. Hitting the bag decreases the volume and increases the pressure to a point where the bag ruptures.

Drinking Through a Straw: Drinking through a straw involves reducing the pressure inside your mouth so that the atmospheric pressure can push liquid up the straw and into your mouth.

Collapsing Juice Boxes: When drinking liquid from a juice box with a straw, the volume of air inside the juice box increases as the liquid is consumed. This decreases the pressure inside the juice box and the box begins to collapse under the atmospheric pressure.

Gas and Scuba Tanks: Storage tanks that hold lots of gas (hydrogen, nitrogen, carbon dioxide, air in scuba tanks, etc.) are designed to be very strong. The number of molecules of gas inside the tank is very large; consequently, so is the pressure inside the tank. The tanks must be strong and thick-walled to keep the gas pressure from exploding the tank.

Bicycle Pump: A bicycle pump adds air to a tire by using a plunger to compress a volume of air inside the pump. The decrease in volume increases the pressure in the pump above the tire pressure and moves more air into the tire. When the plunger is pulled out, a value closes off the tire from the pump so the air will not come rushing back out of the tire and into the pump. This cycle in repeated until the tire is fully pressurized.

Vacuum Pump: A vacuum pump is a bicycle pump in reverse. Consider a pump connected to a chamber of air that is to be evacuated. A plunger in the pump is expanded to increase the volume inside the pump. This lowers the pressure inside the pump, and, as a result, some of the air inside the chamber moves into the pump. A valve allows this air to be removed from the pump when the plunger is compressed. The cycle is repeated.

Taking Blood: When a doctor or nurse takes blood from your arm, she or he pulls a plunger slowly out from a syringe. This increase in volume reduces the pressure in the syringe and the blood, at higher pressure in your arm, enters the syringe.

Pressure Change With Altitude: The atmospheric pressure is largest at the surface of the earth. This is because both the temperature and the density of the air are largest at the surface (largest T and largest N). As you go up in altitude, both the temperature and density decrease and, consequently, so does the pressure.

Space Suits: Astronauts must wear space suits when they are outside the space shuttle during a space walk. Since there is no air in space, there is no air pressure. The space suits provide an artificial air pressure (as well as air to breath) on the body similar to atmospheric pressure at the surface of the earth.

Bubble Domes: Air pressure can be used to support a roof structure. Light domes (usually made of plastic) can be held up by air pressure inside the building. Air is pumped into the building (continually, because of leaks) to keep the inside pressure larger than the atmospheric pressure outside. The pressure difference can support the roof. If the inside air is hotter than the outside air, as it would be in the winter, the gas pressure inside is also increased because of the higher temperature.

Airplane and Mountain Travel: The pressure inside an airplane when it is flying high above the ground is somewhat less than the pressure inside the airplane when it is on the ground. As a result, packages that are filled and sealed at ground level will be expanded when the plane reaches its cruising altitude. The same is true of sealed packages that are taken up a mountain to higher altitudes. Your eardrums "pop" when you ascend or descend in an airplane or drive up and down a mountain. As you ascend and the air pressure is decreased, each eardrum feels a difference in air pressure between the inside of the eardrum (which is still at its original pressure) and air pressure outside of the eardrum (which feels the decreased pressure). To continue to equalize these pressures, you must open the mouth cavity (yawn or chew gum) to bring the pressure inside of the ear to the same value it has outside the ear. The same is true when you descend. When young babies are flying in a plane or traveling in the mountains, they often cry because of the buildup of this pressure difference in the ears—which they have not yet learned how to equalize.

Ascending Balloons: A helium balloon grows in size (and may rupture) as it goes higher into the atmosphere. This is a result of the fact that the balloon experiences an atmospheric pressure that is decreasing with elevation.

Heimlich Maneuver: When a person is choking on a piece of food that has been caught in the airway to the lungs, you can perform the Heimlich maneuver in order to dislodge the food. Approach the person from behind and place both arms around the person just below the rib cage and give a quick and hard squeeze. Doing this will decrease the volume of air in the lungs and increase the pressure. This increase in pressure can dislodge the trapped food.

Warnings on Cans: Many cans that contain gases have a warning label saying not to place them too near a heat source or leave them in the sun. If the gas inside the can gets too hot, the pressure in the can may become so large as to explode the can.

The Freezer Door: Opening a freezer door often takes a little extra pull. The cold air inside the freezer is at a lesser pressure than the hotter air outside the freezer door. This pressure difference makes it harder to open the door.

Dent in a Ping-Pong Ball: To get a dent out of a Ping-Pong ball (or any table-tennis ball), place the ball in boiling water. This will increase the temperature of air inside the ball and the pressure will increase. The dent will pop out.

Basketball in Winter: If you leave a basketball or any other type of air-inflated ball outside overnight in the winter, the ball will not bounce very well in the morning. When the air gets cold inside the ball, the pressure inside the ball is decreased. As soon as the ball warms up again, it will bounce normally.

Suction Cups and Plumber's Helpers: By pushing down on a suction cup (or plumber's helper), the air in the space between the suction cup (or plumber's helper) and the surface is partly removed. This reduces the number of molecules. When the suction cup springs back (or when you pull back on the plumber's helper), the volume of trapped air is increased. Both of these factors (smaller N and larger V) *reduce* the pressure. Since the large atmospheric pressure outside the cup (or helper) has not changed, the pressure difference is plenty to hold the cup (or helper) to the surface.

Eyedroppers, Turkey Basters, and Squirt Guns: First you squeeze the rubber part of the eyedropper. Then you insert the tip (the open end) in a liquid. When you release the bulb, liquid will rise up into the eyedropper. When the bulb is released, the volume of the air trapped inside of the eyedropper increases. This lowers the pressure inside the dropper. The atmospheric pressure pushing down on the surface of the liquid can now push the liquid into the dropper. To get the liquid out of the dropper, you squeeze the bulb and decrease the volume. This increases the pressure and pushes the liquid out. The same principle applies to turkey basters and squirt guns, as well as to a number of other devices that are used to pick up and squirt liquids.

Sticking Cups and Buckets: After stacking cups or buckets one inside the other, trying to pull them apart can be difficult. As you begin to pull, you increase the volume of air between the top cup or bucket and the next. This reduces the pressure in that volume of air. The pressure difference between the atmosphere and this reduced pressure can momentarily "stick" the cups or buckets together.

Lightbulb Implosion: When you accidentally drop a lightbulb (or fluorescent light tube) and it breaks, you often hear a popping sound. This is a result of the fact that the bulb implodes when it breaks. The gas pressure inside a lightbulb is held at a smaller pressure than the atmospheric pressure that surrounds the bulb. This pressure difference causes the implosion. The same is true of the cathode ray tubes inside television sets and computer monitors.

Heat Engines: Many engines run on combustion in a chamber that produces a high temperature gas. The large pressures associated with these hot gases are used to push pistons and drive the engine.

Barometers: These devices are used to measure the pressure of the atmosphere.

LIQUID PRESSURE

Concepts

Unlike gases, liquids are nearly *incompressible.* In other words, it is easy to compress or expand a gas (change its density), but not so for a liquid. If you attempt to change the volume of a sample of liquid, say, by placing it under high pressure, it will resist the compression. Even when subjected to high pressures, the density of a liquid remains about the same. Unlike gases, where the molecules that make up the gas are far apart (and, because of this, can be easily compressed), the molecules in a liquid are essentially touching each other; consequently, they cannot be compressed much more into each other. Because of this incompressibility, liquids are often characterized by their unique densities.

Unlike gases, which tend to fill up the entire volume of the container they occupy, liquids fill a container only up to a given level, a level that depends on the size and shape of the container. This is true of liquids placed in a container near the surface of the earth (or other planets) under the gravitational pull of the planet. It is interesting to note that liquids that are not in a gravitational field of a planet (say, out in space) or for liquids falling to a planet (e.g., raindrops falling to the earth or water inside a space shuttle "falling" in earth orbit), will not fill a container but will cohere into a spherical drop of liquid.

As we have seen, the pressure of a gas in a container depends on three factors: the number of molecules (N) that make up the gas, the temperature (T) of the gas, and the volume (V) occupied by the gas. Now we need to ask the question, "What factors determine the pressure in a liquid in a container (say, water in an aquarium) near the surface of the earth?"

What first comes to mind—based on our personal experience with diving into a swimming pool—is that the pressure must increase with depth. The deeper you are in a liquid, the larger the pressure. Indeed, the pressure (P) grows *directly with the depth* (H) *below the surface.* The reason for this is quite simple. An object placed in a liquid at some depth below the surface feels a pressure due to the entire weight of the liquid column above it. Consequently, the deeper the object is in the liquid, the more weight of liquid above it, and the higher the pressure. It is important to note that this does not mean that the pressure is only downward. At a given depth, the pressure is the *same in all directions* (up, down, sideways, etc.). For example, if you were 10 feet deep in a swimming pool, you would feel the same pressure on an eardrum whatever the orientation. Your eardrum could be facing upward, facing sideways, or facing downward at that depth. The pressure would be the same in all cases. The pressure is said to be *isotropic*—the same in all directions.

A second factor that determines the pressure at a given depth in a liquid is the type of liquid being considered (freshwater, saltwater, oil, alcohol, mercury, etc.). In fact, the pressure (P) depends *directly on the density* (D) *of the liquid.* At a given depth, say, in saltwater, which is denser than freshwater, the pressure would be higher than it would be at the same depth in freshwater. This makes sense because a more dense liquid would give rise to a heavier column of liquid above an object placed in the liquid. It is important to note that the temperature of the liquid can affect this density dependence. In general, hotter liquids are less dense than colder liquids, because the increased thermal motion of the molecules tends to keep the molecules farther apart.

A third factor that determines the pressure (P) at a given depth in a given liquid is *the pressure being exerted on the surface of the liquid,* called *the surface pressure* (Ps). The surface pressure is often the pressure of the gas above the liquid, and, most often, atmospheric pressure, but it can also be a mechanical pressure provided by, say, a plunger pushing directly on the surface. The fact that the pressure at a given depth depends on the surface pressure, often called *Pascal's Principle* (Blaise Pascal, 1623–1662), makes sense because this extra downward pressure can be considered as an addition to the weight of the column of liquid above an object placed in the liquid. Actually, Pascal's Principle is more general. External pressure applied at any place on or in a liquid (not just on the surface) is transmitted equally to all parts of the liquid. For example, if you squeeze a liquid from the side, like when you squeeze a water bottle, the increase in pressure will be felt throughout the liquid.

The fourth and last factor that determines the pressure (P) at a given depth in a given liquid with a given surface pressure is *the strength of the gravitational field* (g) *of the planet.* This makes sense because the gravitational field is responsible for the weight of the liquid. Planets with more gravity—larger gravitational fields—would produce higher pressures at a given depth in a given liquid with a given surface pressure. For example, you would experience a lower pressure at a given depth in a swimming pool on the moon (even without a change in the surface pressure) because the moon's gravitational field is smaller than the earth's field.

In summary, the pressure in a liquid (P) near the surface of a planet (like Earth) depends on only four factors associated with the liquid and its surroundings: the depth below the surface of the liquid (H), the density of the liquid (D), the pressure on the surface of the liquid (Ps), and the strength of the gravitational field in which the liquid sits (g). We can say that P grows linearly with H, D, Ps, and g. In algebraic form, we can express these relationships in a single mathematical expression:

$$P = Ps + DgH$$

This quantitative relationship works quite well for incompressible liquids at rest (not flowing) near the surface of a planet.

A couple of closing remarks are in order. First, it is surprising to note that the pressure at a given depth in a liquid *does not depend on the shape or size of the container.* For example, the pressure is the same 10 feet deep in a large freshwater lake as it is 10 feet deep in a small swimming pool. Second, in a situation where gravity is not a factor—out in space or when a liquid is falling—the pressure inside the liquid (which coheres into a spherical shape) depends *only* on the surface pressure, that is, only on the pressure surrounding the sphere of liquid.

Activities

Book Pile Analogy: You will need to stack a number of books on top of each other. Large identical books, like encyclopedias, work well. In direct analogy to how the pressure increases with depth in a liquid, place your fingers between two books at some point in the stack and feel the pressure. Now place your fingers between two books that are much "deeper" in the stack (larger H) and feel the higher pressure. While there is no change in the density with depth (books do not compress much), the pressure (related to the weight of books above your fingers) increases with depth. To simulate the dependence on surface pressure, repeat with someone pushing down on the top book (increased Ps). To simulate the dependence on density (D), stack objects of different densities (empty boxes, filled boxes, etc).

Homemade Pressure Gauge: You will need to locate a funnel, a length of plastic tubing, a balloon, food coloring, and an aquarium. Take the piece of tubing (6 feet long or so) and attach it to a wall (duct tape) in the shape of a U. Partially fill the U-tube with colored water. The water should be at the bottom of the U-tube and partially up both sides. Attach the funnel to one end of the tubing. The funnel needs to fit tightly into the tubing. Cut a balloon in half and stretch it over the funnel to form an airtight diaphragm. Push gently on the balloon-diaphragm and observe that the water in the U-tube moves. Pull gently on the diaphragm and observe the water in the U-tube move again, but, in this case, in the opposite direction. The difference in the height of the water in the U-tube is a measure of the pressure being exerted on the diaphragm. Now place the diaphragm in an aquarium filled with water and notice the difference in height of the water in the U-tube as you move the funnel deeper and deeper into the aquarium. Pressure increases with depth (H). At a given depth, orient the diaphragm in different directions to see if the pressure is isotropic (same in all directions).

Hole in a Bottle With Pump: You will need to locate a large plastic soda bottle and purchase a thumb pump (Pressure Pumper, Arbor Scientific) that screws onto the top. Punch some holes in the soda bottle at various locations. A good way to punch holes in the plastic is to use a nail whose tip has been heated with a match. The hot nail will melt the plastic and produce a round hole. Over a sink, fill the bottle with water and quickly screw the pump on the top. Increase the pressure of the air above the water surface (increase Ps) in the bottle by pumping on the pump. Notice that all the streams of water exit more quickly. Pascal's Principle: the pressure at the surface is transmitted to all points in the liquid.

Vertical Holes in a Bottle: You will need to locate a large plastic soda bottle and a sink. Punch three holes in the soda bottle at three different heights (one near the bottom, one in the middle, and one near the top). Over a sink, fill the bottle with water and observe the streams of water as they exit the three holes. The stream at the bottom will exit with higher speed (more pressure with increase in depth H). The stream in the middle will exit with medium speed, and the stream at the top will exit with the smallest speed. To illustrate the dependence on surface pressure, place your mouth over the bottle opening and blow some air into bottle (increase N) to increase the pressure on the surface. Note that all of the streams now exit the bottle more quickly. Now try sucking some air out of the bottle (decrease N) and note that all the streams exit the bottle less quickly.

Syringe With Water and Air: You will need to locate or purchase a plastic syringe with cap (Disposable Syringes, Carolina). Cap the syringe. Fill the syringe with water and place the plunger on top of the water. Try to compress the water. Now pour the water out and replace the plunger. Try to compress the air. Water (liquid) is incompressible. Air (gas) can be compressed. Fill the syringe again with water and place the plunger on top of the water. Remove the cap. Push with the plunger on the surface of the water and notice how quickly the water exits the syringe. With more surface pressure (Ps) provided by the plunger, the more pressure there is in the liquid and the faster water leaves the syringe.

Syringes With Water: You will need to locate two plastic syringes and a piece of plastic tubing. Connect a length of plastic tubing between the tips of the two syringes (the connections must be water tight) and fill the system (two syringes and tubing) with water. Place the plungers in each of the syringes and try to remove all air from above the water in each. Push on one plunger and notice how easily the other moves. Pressure on the surface (Ps) is transmitted to the water and pushes the other plunger out. This is the basis of all hydraulic systems (e.g., hydraulic brakes).

Inverted Glass in Water: You will need an aquarium filled with water and an empty glass. Turn the glass upside down and push it straight down into the aquarium of water. Notice that the volume of trapped air in the glass gets smaller (V smaller, P larger) as you push the glass deeper and deeper in the water. Pressure increases with depth (H).

Everyday Examples

Diving: Your eardrums feel the pressure increase as you descend in water. Eardrums can be ruptured in some cases.

Eyes Popping in Fish: If you a catch a fish deep in the ocean and quickly bring it to the surface, the fish might not survive the large pressure decrease. In some cases, you find that the eyes have bulged out of their sockets because of the large reduction in pressure as the fish comes up from the depths of the ocean.

Rising Bubbles: Air bubbles rising in water get larger and larger as they ascend due to the decrease in pressure as they approach the surface.

Dam Structure: Dams are constructed with more material at the base because the pressure in a body of water behind the dam increases with depth.

Water Tank Placement: Community water tanks are placed on the top of a tower or mountain above the town in order to provide water under pressure for homes and businesses. Pressure increases with depth. In cities, you often find water tanks on the tops of the high-rise buildings to provide the needed water pressure. In these cases, the water pressure on the lower floors is higher than that on the top floors.

Submarines: Submarines and other vessels that descend deep in the ocean are designed with very thick metal bodies to withstand the large water pressures deep in the ocean.

Swimming Masks: When you view a scuba diver deep in water you can see the swimming mask pushed hard into the face. When the mask is removed, a ring of pressure can often be seen on the face of the diver.

Fish Sensitivity: Fish are sensitive to pressure changes in the atmosphere. Changes in surface pressure will change the pressure in the water.

Crushing a Styrofoam Cup: For fun, oceanographers often send a Styrofoam cup down with their equipment to the depths of the oceans. When the cup is brought back to the surface with the equipment, it has been compressed to a very small size.

Deflating an Air Mattress: A clever way of getting air out of an air mattress is to submerge the air mattress in water (lake) and use the water pressure to push the air out of the mattress.

Water Tanks and Grain Silos: Large tanks that store water often have more densely packed reinforcing rings near the bottom of the tank compared to the top because the pressure increases with depth. You can also see this reinforcing structure in some grain silos—the grain acts like a fluid, with pressure increasing with depth.

Pump Bottles and Squirt Guns: Pump bottles and squirt guns usually allow you to push a pump that pressurizes (increase the pressure of) the air above the liquid. This pressure on the surface is transmitted to the liquid and causes the liquid to exit the bottle or gun (usually through a tube).

Snorkels, Lungs, and Scuba Regulators: When you are swimming just a few feet below the surface of water, the pressure on your chest cavity (and hence your lungs) is large enough to make breathing through a snorkel difficult. Snorkels work only if you are close to the water surface. When you dive deeper, you need to breathe air that is pressurized to match the water pressure outside your chest cavity. Regulators on scuba gear are designed to do just that. They continually change the pressure of the tank air entering your lungs to match the outside water pressure—so you can breathe easily.

Water Rockets: Toy water rockets require that you use a hand pump to pressurize the air in a chamber above the water. When you release the rocket, the air pressure on the water ejects the water from the rocket and the rocket soars high into the air.

Hydraulic Systems: All hydraulic systems are based on the principle that a change in pressure on the surface of (or anywhere on or inside) a liquid is transmitted to all parts of the liquid.

Intravenous Feeding: The bags of liquids that contain the medicine and food supplements for hospital patients are always hung on a pole above the patient. Liquid pressure increases with depth, and this pressure will force the liquid into the patient's bloodstream.

ARCHIMEDES' PRINCIPLE: SINKING AND FLOATING

Concepts

The fact that pressure increases with depth in a nonflowing liquid near the surface of the earth has an interesting consequence for objects that are placed in the liquid. Consider an object (say, a ball) totally sub-merged in a container of water. Since the pressure increases with depth in the water, the bottom parts of the ball (deepest in the water) will feel larger upward pressures than the downward pressures felt by the top part of the ball (which is not as deep in the water). The overall effect on the ball is an upward force. This upward force is called the *buoyant force* on the object. You can feel this upward force when you attempt to push a basketball underwater. The buoyant force is not a new force in nature. It is simply a consequence of the pressure increasing with depth in the liquid. It is important to realize that the buoyant force on the ball cannot depend on the specific material that makes up the ball. It does not matter whether the ball is made out of steel or foam. As long as the balls are of the same size and both are totally submerged in the liquid, the buoyant force will be the same on both. This must be true because the pressure forces responsible for the buoyant force are arising purely from the external pressure the liquid exerts, which has nothing to do with the type of material that makes up the ball.

So why is it, then, when you submerge a steel ball in water and release it, it quickly descends to bottom, but when you submerge and release a foam ball of the same size, it shoots to the surface? The answer can be found when you consider the *net force* on the ball. It is the net force on an object that causes it to move—to accelerate. Besides the upward buoyant force on the steel ball, there is another force acting on the ball, namely the downward force of gravity (the "weight" of the ball). The downward weight of the steel ball is larger than the upward buoyant force, so the *net force will be downward* and the ball will fall to the bottom. On the other hand, the downward weight of the foam ball is less than the upward buoyant force, so the *net force will be upward* and the ball will rise to the surface. In general, the buoyant force on an object placed in a liquid is never the only force on the object. The weight of the object must always be considered. In fact, it is the tug-of-war between the upward buoyant force and the downward weight that determines the motion of the object. If the weight is larger or smaller than the buoyant force, then the object will sink or rise, respectively. If the tug-of-war happens to be a draw, the object (said to be "neutrally buoyant") will stay suspended within the liquid. It will "flink."

In order to predict whether an object will sink, float, or flink, we need to know how the weight and buoyant force compare. We know how to determine the weight of an object: put the object on a scale and measure its weight. But how can we determine the strength of the buoyant force on an object? Archimedes (3rd century BC) answered this question through some very clever reasoning. Let's reconsider the steel and foam balls, both of the same size and totally submerged in a liquid. We argued that both balls would feel the same buoyant force, but since their weights are different, the steel ball will sink and the foam ball will rise. Now imagine a third ball, identical in size to the steel and foam balls, but made out of the liquid, yes, the same liquid in which it and the other two balls are submerged. The steel, foam, and liquid balls all experience the same buoyant force. Now imagine releasing the liquid ball. What will it do? Of course, it flinks! We must conclude that the buoyant force on the liquid ball is identical to the weight of the liquid ball. But this is the same buoyant force that is being exerted on the steel and foam balls. Eureka! We can only conclude that the buoyant force on the steel and foam balls must equal the weight of this liquid ball. Since this liquid ball has the same shape as the other balls, the weight of the liquid ball is identical to the weight of the liquid *displaced* by the other balls. We have found the answer to our question. Known as Archimedes' Principle, *the buoyant force on an object is always equal to the weight of the fluid displaced by the object.* It works for both gases (like the atmosphere) and liquids (like water) near the surface of a planet (like Earth). Indeed, an object in the atmosphere feels a buoyant force equal to the weight of the air displaced by the object. An object in a liquid feels a buoyant force equal to the weight of the liquid displaced by the object.

A totally submerged object will *sink* if its weight is larger than the weight of the fluid it displaces.

A totally submerged object will *rise* if its weight is less than the weight of the fluid it displaces.

A totally submerged object will *flink* if its weight is exactly equal to the weight of the fluid it displaces.

The above three cases can also be interpreted in terms of *density* (weight or mass per volume). Since a totally submerged object displaces an amount of fluid equal to its volume, we can conclude that objects denser than the fluid will sink, less dense than the fluid will rise, and of equal density will flink.

An object *floating* at the surface of a liquid is not totally but only partially submerged in the liquid. Since the buoyant force still equals the weight of displaced fluid, the floating object needs to displace only a volume of liquid whose weight is exactly equal to its own weight.

Activities

Aluminum Ball and Pancake: You will need to locate some aluminum foil, a hammer, and a container of water. Cut two identical sheets of aluminum foil (same size). Take one of these sheets and crush it with your hands to make an aluminum ball. Take the other sheet and again crush it into a ball but this time pound the aluminum ball with a hammer until it is as flat as a pancake. Both balls weigh the same, but the aluminum pancake is much more dense than the aluminum ball. Place both the ball and the pancake in a container of water. Notice that the less dense aluminum ball (same weight but larger volume) floats, but the more dense aluminum pancake (same weight but smaller volume) sinks.

Wood Blocks: You will need to buy pieces of different types of wood (cork, pine, maple, mahogany, oak, ebony, cherry, etc.) and cut each into identical rectangular blocks. Weigh each block on a scale. Observe how deep each block floats in water. The heavier blocks must displace a larger volume of water to produce a larger buoyant force to support their larger weight. In terms of density, the denser blocks float lower in the water. It helps to waterproof the wood blocks with varnish.

Sinking Wood and Floating Rocks: You will need to buy some special wood (Ironwood, Educational Innovations) that sinks in water and a rock (Giant Pumice, Educational Innovations) that floats on water. The ironwood is denser than water. The rock is less dense than water.

Fake Bricks and Rocks: You will need to purchase a fake brick (www.magic legends.com/product749.html) and a fake rock (www.magiclegends .com/product1087.html), both made out of foam. The foam is shaped and painted to look like a brick and a rock. Place them in water and observe how high they float.

Boats: You will need to locate some oil-based clay (Roma Plastilina, www .dickblick.com/categories/modelingclays/), aluminum foil, a potato, a yogurt lid, and some metal washers. Make boats out of the clay, the aluminum foil, and the potato that will float in room-temperature water. While these materials are all denser than water and usually sink, when formed into the shape of a boat, they can displace more water and float. Add the weights (metal washers) to your boats to see how much weight each can hold. See how many weights a yogurt lid can hold before sinking.

Make a Flinker: You will need to put together a tub full of a variety of objects (corks, balloons, metal washers, putty, paper clips, Popsicle sticks, pieces of wood, etc.). You will also need to locate an aquarium or large container of water. The challenge is to make an object (through trial and error) out of these materials that will neither sink nor float but will remain suspended in the water—a flinker. This is an engaging and challenging exercise. Students will soon discover that even the slightest difference between the weight and buoyant force will either take the object to the surface or take it to the bottom. Another activity that could be used as a warm-up to this activity (using these materials) would be to challenge students to make an object that floats sink and to make an object that sinks float.

Patterns in Density: You will need to locate triplets of objects of the same density but different sizes (three different sizes of corks, three different sizes of glass marbles, three different sizes of steel balls, three different sizes of wood balls, etc.). Place these triplets in water and see what happens. Notice that a family of objects either floats or sinks, with the denser families sinking and the less dense families floating.

Diet Versus Regular Soda: You will need to purchase unopened cans of soda and diet soda (Coke and Diet Coke, Pepsi and Diet Pepsi, etc.). Place the cans in room-temperature water to see which ones sink and which ones float. You will discover that the regular sodas sink and diet sodas float. Weigh each can. Since the volumes of the cans are all equal, the heavier cans are also the denser. It turns out that regular soda (contents plus can) is slightly denser than water, so it sinks. Diet soda (contents plus can) is slightly less dense than water, so it floats.

Saltwater Versus Freshwater: You will need to locate two large containers of water and a box of salt. Prepare one container with saltwater. Use lots of salt—until the water becomes saturated. Prepare the other container with freshwater. See if you can find objects that will float in the salt solution and sink in the freshwater. For example, an egg and regular soda will float in saltwater but sink in fresh. Can you find other objects that do this? Saltwater is denser than freshwater. Since the buoyant force equals the weight of the displaced liquid, objects do not have to displace as much liquid in order to float in saltwater compared to fresh. Objects that are denser than freshwater but less dense than saltwater (like an egg and regular soda) will float in saltwater and not in fresh.

Balls of Clay: Make different sized clay balls out of an oil-based clay (Roma Plastilina, www.dickblick.com/categories/modelingclays/). Observe that no matter what the size of the ball, it sinks in water. All of the clay balls have the same density. They all sink because this density is larger than the density of water.

Layering Liquids: You will need to locate some clear plastic straws, potatoes, food coloring, kosher salt, and eyedroppers. Prepare the following four separate salt solutions:

2 cups of water with no salt and 25 drops of blue food coloring

2 cups of water with 4 tablespoons of salt and 25 drops of red food coloring

2 cups of water with 8 tablespoons of salt and 25 drops of yellow food coloring

2 cups of water with 14 tablespoons of salt and 25 drops of green food coloring

Have the students attempt to layer these liquids in a straw (by trial and error), using an eyedropper to drop the liquids into the straw. The straw should be cut in half (half straw) with one end inserted at a 45-degree angle into a slice of potato (makes a great holder). You can also use a piece of clay as the straw holder instead of the potato slice. Through trial and error, the students will discover the density ordering from bottom to top (green, yellow, red, blue). Students may want to try other substances that dissolve in water besides salt—such as sugar or baking soda. Students may want to try other liquids. Some liquids that are less dense than water are alcohol, cooking oil, linseed oil, and motor oil. Some liquids that at are denser than water are glycerin and corn syrup. Students may want to try to layer different types of milk (from low fat to full cream).

The Hydrometer: You will need to locate some plastic soda straws, oil-based clay (Roma Plastilina, www.dickblick.com/categories/modelingclays/), three liquids (alcohol, water, glycerin), and containers. You can make a simple hydrometer by placing a small wad of clay at the end of a straw. Place the hydrometer in each of the three liquids. There should be enough clay to float the straw partway down in the liquids and to float it vertically. Notice that the hydrometer floats highest in the densest liquid (glycerin), not as high in the water, and lowest in the least dense liquid (alcohol). Hydrometers are used to measure the density of a liquid based on this effect. Test your hydrometer in other liquids (see "Layering Liquids," above). You can make a scale on your hydrometer. Mark the level for fresh-water at room temperature at zero. Here is another thing to try. If you push the hydrometer down into the water a little below its floating position and let it go, it will bob up and down for a few cycles. Pushing the hydrometer down increases the buoyant force (more liquid displaced). Now the buoyant force is greater than its weight (it was balanced before) and the hydrometer accelerates up. Its inertia carries it above its original floating position. Now the buoyant force is smaller than the weight and it accelerates down. This scenario is repeated for a few cycles before the hydrometer settles back (due to viscous drag) to its original position.

Hot Versus Cold Water: You will need to prepare a container of hot water (hot tap water will do) and a container of cold water (ice water will do). See if you can find objects that will sink in the less dense hot water but float in the denser cold water. Eggs? Cans of soda? Use the simple hydrometer described in "The Hydrometer," above, and see if you can observe the density difference.

Sea of Beans and Turning Black Into White: You will need to locate a Ping-Pong ball and a heavy steel ball of approximately the same radius. Paint the steel ball black. You will also need a large clear bowl filled halfway with dried pinto beans. Hide the Ping-Pong ball in the center of the beans just below the surface. Place the black steel ball on the surface of the beans right above the hidden Ping-Pong ball. Grab the bowl with two hands and swirl the beans. You will "turn" the black ball into a white ball. This is a very engaging "trick" based on the analogy that the steel is denser than the beans and the Ping-Pong ball less dense. The steel sinks into the beans and the Ping-Pong ball rises to the surface.

Cartesian Divers: You will need to locate glass (not plastic) eyedroppers and empty plastic soda bottles with screw tops. You can also purchase very inexpensive kits to make these "Cartesian Divers" (Cartesian Divers, Educational Innovations). Fill the eyedropper with enough water to make it barely float in water. That is, fill the eyedropper with a water-air mixture so that, when you place it in water, its rubber top is just poking through the surface. Fill up a plastic soda bottle with water (leave an inch or so of air near the surface) and place this barely floating eyedropper in it. Screw on the top. Squeeze the bottle with your hands and observe the eyedropper (the diver) sink to the bottom. Release your hands and observe the diver ascend to the surface. When you squeeze on the bottle, the pressure is transferred to all points of the liquid (Pascal's Principle). This increase in pressure forces some additional water into the eyedropper. This increases the weight of the diver enough so that the weight is now larger than the buoyant force. The diver descends. Releasing the bottle releases the pressure, and the water exits the eyedropper. Now the buoyant force is again larger than the weight and the eyedropper ascends. Instead of squeezing the bottle, you can also take the top off and pressurize the air above the surface by blowing air (using your mouth) into the top of the bottle. You can also use a thumb pump (see "Hole in a Bottle With Pump," above) to pressurize the air on top of the water. You can also stretch a balloon diaphragm over the opening of the bottle and push down on it in order to increase the pressure.

Archimedes' Experiment: You will need to purchase a "displacement apparatus," which is nothing more than a cup with a pour spout (Overflow Can and Catch Bucket, Sargent-Welch). You will also need to locate a small spring scale and an object that is denser than water (like a metal cylinder) that you can submerge in the displacement apparatus without touching the sides. Using the spring scale, weigh the object by attaching it directly to the spring scale and lifting it off the ground. Record the weight of the object. Fill the displacement apparatus with water to the point where it begins to spill out the spout. The best way to do this is to use a cup to catch the water until it stops draining out. Now use the spring scale to lower the object slowly and carefully into the water and catch the displaced water in an *empty cup that you have pre-weighed on the balance.* When the object is submerged, record the spring scale reading. See if the difference in the spring scale readings equals the weight of the displaced fluid.

Everyday Examples

Steel Boats: A hunk of steel sinks, but steel in the shape of a boat floats. A hunk of steel cannot displace enough water to make the buoyant force (the weight of the displaced water) as large as its weight. But when the steel is shaped into a boat, it can displace enough water so that the buoyant force matches the weight.

Life Vests and Float Toys: Wearing a life vest increases your volume significantly without adding much to your weight. This increase in volume will allow you to displace more water and increase the buoyant force. The same is true for air-filled float toys, used to help people float.

Ducks: Ducks and other waterfowl have lightweight and waterproof feathers that give them additional volume with very little increase in weight. So ducks float very high in the water.

Submarines: Submarines can descend and ascend in water by pumping water in or out of their holding tanks. Adding water increases the weight over the buoyant force and the sub descends. Removing water decreases the weight and the sub ascends.

Helium and Hot Air Balloons: A helium balloon is less dense than air because helium atoms weight less than the molecules that comprise the air (nitrogen, oxygen, etc.). Hydrogen also works, but hydrogen is very dangerous (the Hindenburg) since it easily combusts with oxygen. Hotter air is less dense than colder air since the "hotter" air molecules are farther apart. A hot air balloon can ascend by heating the air inside the balloon and descend by allowing it to cool.

Air Bladders in Fish: Fish have an air bladder that they can use to ascend and descend in water. By increasing the volume of the bladder, the fish can become less dense and ascend. By decreasing the volume of the bladder, the fish can increase its density and descend.

Ice Cubes and Icebergs: Ice (frozen water) is less dense than liquid water; consequently, ice cubes and icebergs float in water. Ice cubes will sink in alcohol—a good way to tell if someone has a "strong" drink.

Density Toys: Novelty stores sell a variety of toys that contain two or more colored liquids of different density. The less dense liquids float above the more dense liquids. In some of these toys, an object (like a toy boat) floats at the interface of the two liquids. Slow moving waves can also be made at theses interfaces. In some of these toys, less dense liquid drops can be made to rise through a more dense liquid (lava lamps).

Dead Fish Floating: Dead fish rise to the surface and float. Death results in bloating, so the fish's volume is increased (density decreases) and the buoyant force is increased. A similar situation happens for other animals, including humans.

Walking on Pebbles in Water: It is much more comfortable to walk barefoot on pebbles when you are in the ocean or in a lake than when walking on pebbles on land. The upward buoyant force provided by the water on your body can greatly reduce the force that the pebbles exert on your feet. The deeper you go, the more comfortable it gets.

Hot Drinks and Soup: Hot liquid is less dense than colder liquid. So watch out when drinking hot soup or hot drinks, because the hottest liquid is at the surface.

Cold Water Into Lake: When a cold river empties into a warmer lake it sinks down into the lake (colder water is more dense that hotter water). In a lake, hotter water will reside near the surface while colder water will occupy the bottom.

Floating in the Great Salt Lake: It is much easier to float in the Great Salt Lake than in a freshwater lake because the density of saltwater is larger than freshwater.

Floating in Foamy Water: It is difficult to float or swim in foamy water because foamy water is less dense than regular water.

Ships Sinking in Rivers: There are documented cases of an overloaded ship sinking as it leaves the ocean and enters a freshwater river (Thames, Mississippi). Saltwater is denser than freshwater.

Oil on Water: Oil is less dense than water. This is why you often see motor oil from cars floating on puddles of water, oil from motorboats or oil tankers floating on water, and vegetable oil floating above the vinegar and water in a bottle of salad dressing (shake before using).

Floating: Some people can float on water more easily than others—a result of the differences in body densities. The best way to float is to displace as much water as possible (buoyant force equals the weight of the displaced fluid). To maximize the buoyant force, people float on their backs with their entire body, including the arms, underwater except for a small area around the face. Inhaling and expanding your lungs will increase your volume. This will help you float, but when you exhale, your body will sink more.

Bobbing in Water: You often see objects (wood, buoys, boats, etc.) bobbing in water. If an object is pushed down a little below its floating position (say, by wave or wind action), it will bob up and down for a few cycles. When the object is initially pushed down in the water, the buoyant force is increased (more liquid displaced). With the buoyant force now greater than the weight (it was balanced before), the object accelerates up. Its inertia carries it above its original floating position. Now the buoyant force is smaller than the weight, so it decelerates up and then accelerates down. This scenario is repeated for a few cycles (bobbing) before the object settles back (due to viscous drag) to its original floating position.

Hydrometers: Hydrometers are used to measure the density of liquids for a variety of purposes—from beer making to checking antifreeze in a car radiator (see "The Hydrometer," above).

Galileo's Thermometer: These novelty thermometers are based on the fact that liquids expand and become less dense with increasing temperature. The thermometer contains a sealed glass tube filled with a liquid and containing a series of floating glass balls of different weights (and different densities). These balls are arranged with the densest ball at the bottom and the least dense at the top—with all the balls floating at cold temperatures. As the liquid in the tube warms from the heat in the room, the liquid expands and, at a given temperature, the densest ball sinks in the tube. As the temperature continues to rise, other balls sink in order. A temperature tag on each ball reports the temperature in the room.

Floating Mountains: Mountains "float" on the semi-liquid mantle of the earth.

Mixed Salads: Notice that the denser items in a mixed salad (bell peppers, tomatoes, nuts, etc.) sink in the less dense lettuce, simulating the layering of liquids by density.

Hotter Air Rises and Colder Air Falls: Air in a room is hotter nearer the ceiling than the floor—hot air is less dense than cold air. You can experience hot air rising in many other situations as well—over a stove, above a candle or campfire, over a hot drink, up a chimney, above a radiator, and within the atmosphere—to name only a few. If you are sleeping below a window in the winter, you can feel the cold air (cooled by the cold window) falling down on you. Storage freezers are often designed as chests, with their door (lid) opening on the top—colder air sinks and stays within the freezer.

Convection Currents: The fact that hot air rises in colder air and that hot liquids rise in colder liquids is responsible for a very common phenomenon— convection currents. For example, the sun heats the surface of the earth. The warm land heats up the air near the surface. This hot air rises and is replaced by colder air. This sets up cells of circulating air currents that drive many of our weather patterns. Water is heated in a pot on the stove. The hot water near the bottom on the pot (closer to the burner) warms first, expands, and rises toward the surface. Colder water replaces the rising hot water and convection currents are set up in the pan. Convection is also important in oven cooking, home heating, offshore and onshore breezes, motion of the continental plates, heat transport in the sun, and shimmering air rising above a radiator, to name only a few.

Internal Seiches: In the summer, alpine lakes develop a top layer of warmer and less dense water over a bottom layer of colder and denser water. Internal waves, called "internal seiches," are often observed on the "surface" between these layers.

Bays: Many bays that have freshwater from a river or rivers entering the saltwater of the ocean (e.g., the San Francisco Bay) maintain a layer of less dense freshwater over a denser layer of saltwater.

FLUIDS IN MOTION: BERNOULLI'S PRINCIPLE

Concepts

As you will soon see, fluids behave very differently when flowing compared to when they are not moving (wind compared to air in a room or a river compared to a lake). In many cases, the flow can be quite complex and somewhat unpredictable, leading to the formation of whirlpools and other types of turbulence. Such complex flows will not be considered here. We will look only at "well-behaved" flow (sometimes called "ideal" or "laminar" flow), flow that is steady and nonturbulent. We will also assume that the flowing fluid is incompressible (no density changes in the flow). This is a very good assumption for liquids but only approximate for gases. We will also assume that the flow exits without internal or external friction (nonviscous flow).

Consider a well-behaved fluid flowing steadily through a pipe of varying cross-sectional area. As the pipe narrows (smaller cross-sectional area), the flow speed must increase. As the pipe widens (larger cross-sectional area), the flow speed must decrease. The reason for this behavior is related directly to the assumption that the fluid is incompressible (of constant density). To maintain constant density, the speed must increase to traverse the narrower pipe and decrease when occupying the wider pipe. This can also be argued in terms of flow rate. Since the amount of fluid that passes through a given cross-sectional area per second cannot change along the flow (the same volume that goes in must come out), the smaller cross-sectional areas must have faster flows and larger cross-sectional areas must have slower flows to maintain the same volume rate. This relationship between flow speed and cross-sectional area, called the *continuity principle*, can be expressed mathematically. The product of the flow speed (symbol S) and the cross-sectional area (symbol A) equals the constant flow rate (symbol R): $SA = R$. Since the flow rate R is a constant number, a smaller cross-sectional area A gives a larger flow speed S and vice versa.

What about the pressure in a moving fluid? Our intuition probably fails us here. The pressure in a moving fluid is actually *less* than the pressure in a nonmoving, static fluid. Furthermore, the faster the fluid flows, the lower the fluid pressure becomes. Two situations may help us conceptualize this nonintuitive result. First, consider the tale of the Dutch boy with his finger in a hole at the base of the dike and holding back the water. The large static pressure (due to the water above the hole) will be somewhat "relieved" when the boy pulls his finger out and lets the water flow. Consider a second case: water in a garden hose. When the nozzle is closed off, the static water pressure in the hose is large and any holes in the hose or leaks near the faucet connection will shoot out water. But when the nozzle is opened and the water is allowed to flow out of the hose freely, the leaks stop. This means that the pressure inside the hose when the water is flowing must be less than the static pressure in the hose when the water was not flowing. Our intuition may also be influenced by the confusion between the water pressure *within* the moving fluid (being addressed here) versus the force that a moving fluid can exert on objects when it slams into them (which is not being addressed here). The former can be small while the latter large. Daniel Bernoulli (1700–1782) was the first scientist to understand this phenomenon. The physical principle, often called Bernoulli's Principle, states that *the pressure in a moving fluid is less than the pressure in a fluid at rest, and the pressure decreases with increasing fluid speed.*

Activities

Rising Paper: You will need to locate a piece of notebook paper. Hold the paper horizontally, using your thumbs and forefingers to support the paper at the side edges about one-third of the way from the end nearest you. Allowing the paper to droop downward naturally, blow straight out over the paper just above its curved surface. Observe how the paper moves upward toward the moving air. The moving air above the paper is at a lower pressure than the static air below the paper.

Collapsing House: You will need to locate a 4"-by-5" card. Make a half-inch right-angle fold at each end. Place this little two-walled house on a table. Make sure the walls spread outward a little bit from the vertical. Blow hard *under* the card, between the tabletop and the card, through one of the open sides. Observe the downward movement of the roof. The air moving under the card is at lower pressure than the air above the card, so the card moves downward.

Paper Attraction: You will need to locate two pieces of notebook paper. Fold each about two inches from the shorter edge. While holding the papers at the folded edges, bring them close together in front of your mouth. Blow air directly between the sheets. Observe how the papers come together. They may even vibrate and make noise. The moving air between the papers is at lower pressure than the air outside the paper, so the papers come together. The "continuity principle" is also involved here; since the air is being channeled between the papers, it picks up speed due to the smaller cross-sectional area.

Kissing Light Bulbs: You will need two light bulbs and a piece of string. Tie the socket end of each of the light bulbs to the ends of a two-foot-long piece of string. Holding the middle of the string, allow the bulbs to dangle in the air. You can also hang the system from a table with masking tape. In either case, you need to make sure the two bulbs are hanging at the same height and almost, but not quite, touching each other. Blow your breath hard straight between them and observe (hear) what happens. The bulbs will come together and hit (click). Sustained blowing may produce a double or triple hit (click). You might try repeating this activity using helium-filled balloons attached to the floor or ordinary balloons hanging from the ceiling. In either case, a more delicate blowing technique will be required.

Kissing Soda Cans: You will need to locate two empty soda cans and a dozen wood dowels (round pencils will do). Set the dowels parallel to each other on a smooth table about a half-inch apart. Set the soda cans close to each other on top of the dowels about an inch apart. Now blow hard directly between the cans. Observe how the cans come together, moving toward the moving air. The air is channeled between the cans (think continuity principle) and this fast moving air is lower in pressure than the static air on the other sides of the cans.

Soda Can Car: You will need to locate an empty soda can and a light flatbed cart. Set the soda can on and attach it to the center of the cart (with tape, clay, etc.). Blow air from your mouth or from an air supply (vacuum cleaner in its blowing mode or a good handheld hair dryer) across the front side of the can (the side of the can facing the front of the cart) and watch the cart move toward the moving air. Blow on the other side of the can and watch the cart move backward toward the moving air.

Flying Dime: You will need to locate a dime and a shallow bowl. Place the dime a couple of inches back from the edge of a table and place the bowl on the table about eight inches (20 cm) behind the dime. Bend down and blow hard above and straight over (not under) the dime in the direction of the bowl. A hard enough blow will cause the dime to rise off the table and fly into the bowl. Observe how the dime is picked up off the table by the moving air above it. Blowing over the dime reduces the pressure on the top of the dime. Once the dime pops up into the moving airstream, the stream will push it into the bowl.

Levitating Card: You will need to locate a large spool of thread (the kind with a hole running through it), an index card, and a thumbtack. Cut out a 3"-by-3" square from the index card. Punch a thumbtack through the center of this square. Hold the card up against the bottom of the spool with the thumbtack tip inside the hole. Hold the spool vertically with the card at the bottom. Place your mouth on top of the spool and blow continuously downward through the hole. Just after you start blowing, let go of the card, but keep blowing. Observe that the card will remain up against the spool until the blowing stops. The air blown through the hole must exit through the narrow space between the card and spool. This increases the air speed (continuity principle) and reduces the pressure (Bernoulli's Principle) between the card and spool. The difference in pressure between this fast moving air and the static air below the card is plenty to hold the card up against gravity.

Levitating Water: You will need to locate a small glass, two clear plastic straws, and food coloring. Fill the glass with colored water. Hold one of the clear plastic straws down vertically in the colored water. Using a second straw as a directional blower, blow hard and straight over the top of the straw in the water. Observe that water rises up in the vertical straw. The harder you blow, the higher it rises.

Follow the Stream: You will need to locate a Ping-Pong ball, an aquarium tank, and a beaker. Float the Ping-Pong ball in the aquarium tank that is half-filled with water. From a full beaker, pour water slowly on top of the floating ball. As you continue to pour, move the stream of water slowly. Observe that the ball moves in the direction of the lower pressure pouring (moving) water.

Ping-Pong Ball and Spoon: You will need to locate a sink and faucet, a Ping-Pong ball, a piece of thread, tape, and a spoon. Get some water pouring slowly and smoothly out of the faucet. Tape the Ping-Pong ball to one end of a foot-long piece of thread. While holding the other end of the thread, bring the ball up to the moving water. It will be pulled into the stream and can be made to stay there. A similar activity can be performed using a spoon. Dangling the spoon loosely in your fingertips, place the back of the spoon in the faucet stream. Feel and observe it being pulled into the lower pressure moving water.

The Ascending Egg: You will need to locate a large beaker and a raw egg. Fill the beaker with water and place the egg in it. The egg will sink to the bottom. Now stir the water near the top of the tank using a large circular motion with your finger. If you stir vigorously, the egg will rise off the bottom and move into the swirling water near the top of the beaker. The moving water nearer the top of the beaker is lower in pressure than the more static water nearer the bottom.

Funnel Fun: You will need to locate a Ping-Pong ball and a plastic funnel. Place a Ping-Pong ball inside the plastic funnel. With the opening of the funnel pointed straight up, try to blow the ball out of the funnel. It cannot be done. The harder you blow, the less chance you have. The moving air, as it is channeled between the ball and the lower walls of the funnel, is reduced in pressure. This reduction in pressure compared to the larger pressure on top of the ball keeps the ball in the funnel. Now try this activity with the funnel pointed downward. With a constant air supply, the ball can be held in the funnel against gravity.

Floating Balls: Locate an air blower and/or a handheld blow-dryer and a Styrofoam ball. In fact, any type of air supply that blows reasonably hard and whose air direction can be controlled will work. Direct the airstream upward and fix it in place. Gently place the Styrofoam ball in the airflow and release it when you find the point in the stream where the upward push of the air balances the downward weight of the ball. It should stay suspended in the airstream by itself. The ball does not fall out of the stream because of the difference in pressure between the faster moving air in the middle of the airstream (lower pressure) compared to the slower moving air at the edges of the airstream (higher pressure). Make up games using the handheld blower in which the students have to transport the floating ball across the room and drop it into a basket. Try to suspend different objects in the airstream. There are a variety of children's toys that are designed on a similar principle. The most common type involves a pipe through which air is blown to suspend a small ball. There is also a children's game that uses such suspended balls in a space theme.

Balloon Kites: You will need to locate the same air blower as described above (see "Floating Balls," above) and an inflated balloon tied to a string. Direct and fix the airstream upward. Bring the balloon slowly toward the moving airstream. When the balloon begins to enter the region of moving air, you will feel it being both pushed up by and pulled into the moving air. With a little practice, you will be able to hold the balloon in place by the string, like you are flying a kite.

Toilet Paper Fun: You will need to locate the same air blower as described above (in "Floating Balls") and a half-roll of toilet paper. Direct and fix the airstream upward. Bring the roll of toilet paper slowly toward the moving airstream. Place the pointing finger of each hand inside each side of the roll's opening with the end sheet of the toilet paper directed upward toward the airstream. When the roll begins to enter the region of moving air, you will feel it being both pushed up by and pulled into the moving air. Let the roll spin loosely on your fingers. With a little practice, you can get the toilet paper to unroll completely up and into the moving air.

Everyday Examples

Blowing Out Candles: It is difficult to blow out candles with your mouth wide open, so you make a smaller opening with your mouth and blow. The speed of the air coming out of your mouth is much greater if the area through which it leaves is smaller (continuity principle).

Rivers: Rivers flow much more quickly through narrow gorges than through wider stretches of river. The smaller cross-sectional area in the gorge produces a faster flow (continuity principle).

Water From a Hose: To get water to travel farther when it exits a hose, you hold your thumb over part of the outlet. As the area at the outlet becomes smaller, the speed of the water becomes larger (continuity principle).

Wind in the City: Wind can be channeled between buildings in a city and create high wind areas.

Mary Poppins: Umbrellas seem to float on breezy days. The wind moves faster over the top of the umbrella (lower pressure) than underneath (high pressure), and the pressure difference gives lift to the umbrella.

Airplane Wings and Frisbees: The wings of an airplane and the shape of a Frisbee are designed to make the air flow over the top faster (lower pressure) than under the bottom (higher pressure).

Passing Boats and Cars: When moving boats or cars pass each other, they tend to be drawn together. This is a result of the faster flow of fluid (water for boats, air for cars) channeled between them (continuity principle) compared to the outside flow.

The Shower Curtain: The shower curtain moves in to grab your leg. The air in the shower stall, being pushed around by the water coming out of the showerhead, is moving faster than the air outside the curtain. This means that the pressure inside is less than outside and the pressure difference moves the curtain inward.

Convertible Tops and Motorcyclist's Jackets: The outward bulge of a convertible top and the billowing of a motorcyclist's jacket result from the fast moving (and lower pressure) air going over the convertible top and around the jacket, respectively.

Tornadoes: Roofs are blown off (upward) by the high-speed winds of a tornado. Since the air is moving much faster over the top of the roof than inside the house, the pressure inside the house is larger than above the roof. This pressure difference can blow the roof off.

Windows in Building: Windows can be blown out of tall buildings during high winds. Wind is channeled in between buildings, so the air outside the window is moving much faster (lower pressure) than the air inside the building.

Open Windows and Long Hair: Loose papers often fly out of open car windows. You often see long hair being pulled out a car window. The pressure in the moving air just outside the open window is lower than the pressure inside the car.

Carburetors and Atomizers: Carburetors and atomizers (such as perfume or pesticide spray bottles) use air moving quickly through a small tube to draw up liquids.

Chimney Smoke: Smoke will vent through a chimney to the outside of the house because of the air blowing (lower pressure) over the top of the chimney opening. Indeed, a chimney draws better on a windy day.

Sailboats: Sails on sailboats are curved (like an airplane wing) in such a way that the wind passes over the curved side at a higher speed (lower pressure) than over the back side (higher pressure). This pressure difference can help propel the boat.

Toys in a Tub: Toys or objects floating in a bathtub will migrate to and stay under the water flowing out of the faucet. Flowing water is at lower pressure than the nonmoving water in the tub.

Prairie Dog Hole Design: Prairie dogs design their burrow system to circulate air through the underground tunnels by employing the Bernoulli concept. One opening is flat to the prairie while another is in a curved dirt mound. Air flowing over the curved dirt mound moves at a faster speed than that flowing over the flat opening. The pressure difference maintains a steady circulation of fresh air underground.

Vocal Cords and Squealing Balloons: Air rushing through the vocal cords produces a lower pressure region between the chords, which along with the tension in the folds, helps to close them. But soon the pressure builds up behind the closed chords, enough to reopen them and the cycle repeats. The vocal cords vibrate, open and close, in this fashion. Releasing air out of a balloon through a constricted opening (made by pulling out and stretching the rubber opening) causes an irritating squeal, a high frequency vibration produced by the a mechanism similar to that just described for the vocal cords.

Ship by a Dock: As a ship pulls up to a dock, water can be channeled between the boat and dock. This can reduce the pressure and can drive the ship into the dock. Most docks allow water to flow under them to reduce this effect.

FLUIDS AND BUOYANCY CIRCUS

The following set of ten activities, selected from the activities described in the last three sections, might be used to initiate your unit on fluids and buoyancy. These activities could be set up around the classroom in a circus format. Next to each activity, a simple description of how each activity is to be performed would be displayed, along with a question or questions to be answered by the student in conjunction with performing the activity. Obviously, you will need to rewrite these descriptions and questions to make the language and analysis appropriate for your grade level. It is suggested that students work in pairs or small groups. One option would be to have students perform the activities a few at a time and run the circus over a few days. Another option would be to use some of the activities as teacher demonstrations for whole-class discussion. In any case, students should be encouraged to probe the activities beyond the descriptions and begin to think of additional questions they might want to investigate on their own later in the unit.

1. Wood Blocks
Test the buoyant properties of each of these pieces of wood. Make a drawing that shows what you have found out.

2. Boats
Construct an aluminum foil boat, clay boat, and/or potato boat. How much weight (metal washers) can your boat support before it sinks? What are the factors in your investigation that you think make a difference for flotation? List them.

Float one of the plastic lids in the water. Estimate how many weights (ceramic cylinders) the lid will hold before it sinks. Try it out.

3. Patterns in Density
Provided are sets of three objects of different size but made from the same material. Place each set into the water to see if the set sinks or floats. Which sets sink? Which sets float? Are there any patterns in your observations?

4. Can You Make a Flinker?
See if you can create an object that neither sinks nor floats. If you're successful, this object will merely hover below the surface of the water

without breaking the surface and without resting on the bottom of the container of water. Can you name some objects in the world around you that have this "neutrally buoyant" property?

5. Diet Versus Regular Soda
Test the buoyant properties of each of these cans of soda. Compare each can to each of the other cans. Find a way to tell others what you have found.

6. Cartesian Divers
Squeeze the bottle and watch the "diver." What differences do you see in the diver and in the bottle when the bottle is squeezed? Try to use these observations to explain how you think this toy, called a Cartesian Diver, works. Make some drawings with labels to help you explain your ideas.

7. Air Pressure Investigations
Vacuum System: Place the partially inflated balloon inside the chamber. Attach the clear rubber tube coming from the air pump (looks like a big bicycle pump, except it pumps air out, instead of in) to the outlet at the base of the chamber. Pump the air out of the chamber and watch what happens to the balloon. What is your explanation for this? To place the air back into the chamber, just pull the rubber tube off of the base of the chamber.

Rubber and Soda Can Mats: Take the square rubber mat and place it flat on the tabletop. Using the handle in the middle, try to pull it off the table. What is your explanation for this effect?

Suction Cup Fun: Take the two "suction cups" and squeeze them together. Try to pull them apart. What do you think is happening here?

Sandwich Bag in a Cup: A clear garbage bag has been inserted in the trash can and sealed tightly around the outer edge of the can. Reach into the trash can and grab the bag and try to pull it out slowly. What is your explanation for this effect?

Funnel Fun: An empty soda bottle has been fitted with a funnel and sealed with clay. Pour water into the funnel and observe what happens. Why do you think this happens?

8. Homemade Pressure Gauge
A simple "pressure" gauge has been constructed from a funnel and balloon, connected to a U-shaped tube partially filled with colored water. Gently "push" on the balloon with your finger. Gently "pull" on the balloon with your fingers. In each case, note what happens to the water level in the U-tube.

Now take the funnel and slowly lower it into the large beaker of water. Describe how the water level in the U-tube changes with funnel depth.

Can you name some personal experiences or everyday examples that relate to what you just observed?

9. Archimedes' Experiment

You have been provided with a "displacement apparatus"—a cup with a pour spout. Using the handheld spring scale provided, weigh the metal cylinder by attaching it directly to the spring scale and lifting it off the table. Record the weight of the cylinder. Fill the displacement apparatus with water to the point where it begins to spill out the spout. The best way to do this is to use a cup to catch the water until it stops draining out.

Now use the spring scale to lower the cylinder slowly and carefully into the water and catch the displaced water in an *empty cup that you have pre-weighed on the balance.* When the cylinder is submerged, record the spring scale reading. How much less does the cylinder weigh in water? Also, weigh the cup + displaced water on the balance. What is the weight of the displaced water? Describe any patterns or relationships that you notice.

10. Layering Liquids

Try to layer all four liquids in the straw so that none mix. Find a way to keep track of your results (successful and unsuccessful attempts).

A BERNOULLI CIRCUS

Perform the 10 fun activities described below. All of these activities are based on a single scientific principle. Can you identify what is common to all these activities? The pattern?

1. Rising Paper

Hold the paper horizontally using your thumbs and forefingers to support the paper at the side edges about one-third of the way from the end nearest you. Allowing the paper to droop downward naturally, blow straight out over the paper just above its slightly curved surface. Which way did the paper move? Up or down?

2. Collapsing House

Place the little two-walled house on a table. Make sure the walls spread outward a little bit from the vertical. Blow hard *under* the card, between the tabletop and the ceiling, through one of the open sides. Which way did the paper move? Up or down?

3. Paper Attraction

Blow some air directly between the two sheets of paper. Which way did the papers move? Toward or away from each other?

4. Kissing Light Bulbs

In front of you are two light bulbs that are hanging at the same height and almost but not quite touching each other. Blow hard right between them. Do they come together or move apart?

5. Kissing Soda Cans

Set the soda cans close to each other on top of dowels placed about an inch apart. Blow directly between the two cans. Do they come together or move apart?

6. Levitating Card

Hold the card up against the bottom of the spool with the thumbtack tip inside one of the spool's holes. Hold the spool vertically with the card at the bottom. Place your mouth on top of the spool and blow continuously downward through the hole. Just after you start blowing, let go of the card, but keep blowing. Does the card fall off when you are blowing?

7. Levitating Water

Take one of the clear plastic straws and hold it down vertically in the colored water. Using a second straw as a directional blower, blow straight over the top of the straw in the water. What happens to the water in the straw?

8. Follow the Stream

Float the Ping-Pong ball in the aquarium tank that is half-filled with water. From a full beaker, pour water slowly on top of the floating ball. As you continue to pour, move the stream of water slowly. Does the ball move along with the moving stream of water or not?

9. Ping-Pong Ball and Spoon

While holding the end of the thread, bring the Ping-Pong ball up to the slow-moving water stream from the faucet. Is the ball pulled into the stream or repelled by it? Dangle the spoon loosely in your fingertips. Place the back side of the spoon in the slow-moving faucet stream. Is the spoon pulled into the stream or repelled by it?

10. Funnel Fun

Place a Ping-Pong ball inside the plastic funnel. With the opening of the funnel pointed straight up, try to blow the ball out of the funnel. Can you do it?

SAMPLE INVESTIGABLE QUESTIONS

- *Wood Blocks:* If a given type of wood (say, oak) were cut into different size square blocks (all with the same vertical dimension), would they all float at the same depth in the water?

- *Boats:* Does the *volume* of the aluminum boat (floor area multiplied by the height of the walls) determine how many washers the boat can hold before sinking?

- *Make a Flinker:* How does the density of a flinker (mass divided by volume) compare to the density of water (1.0 grams per cubic centimeter)? How can you determine the volume of a flinker through a water displacement method in order to calculate its density?

- *Diet Versus Regular Soda:* Will a can of regular soda that sinks in freshwater float in saltwater? Will a can of diet soda that floats in regular water sink in alcohol? Will a can of diet soda that floats in water at room temperature sink in hot water?

- *Cartesian Divers:* Can you make a Cartesian diver work in a glass bottle using a rubber diaphragm (balloon) stretched over the opening? Can you make a Cartesian diver out of something else besides an eyedropper?

- *Air Pressure Investigations:* What other objects can you place inside the vacuum system that will expand when air is removed from the chamber?

- *Suction Cup Fun:* Does the size of a suction cup determine how much weight it can support?

- *Homemade Pressure Gauge:* What would a graph of the pressure (difference in height of the water columns in the U-tube) versus the depth in the liquid (distance the diaphragm is placed below the surface) look like? Is the graph a straight line? Would the graph change if saltwater were used instead of freshwater?

- *Layering Liquids:* What other types of liquids will layer in a straw?

- *Rising Paper, Collapsing House, Paper Attraction, Kissing Light Bulbs, Kissing Soda Cans, Levitating Card, and Levitating Water:* Can all these activities be performed using air blown from a handheld hair dryer?

Waves and Sound

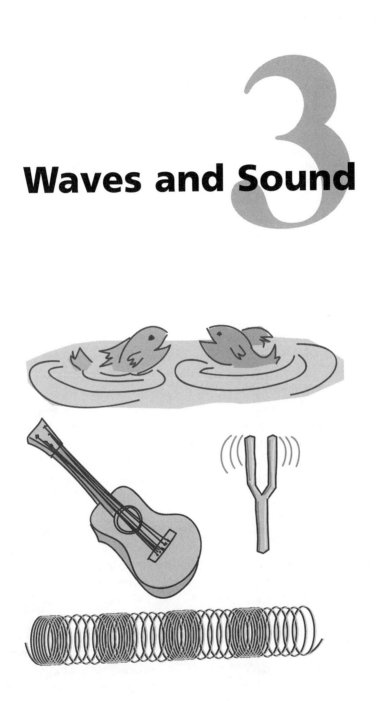

WAVES AND WAVE MOTION

Concepts

Types of Waves and Wave Speed

Energy and information can be produced, sent, and delivered in two distinct ways. One way involves the actual physical transport of matter from one place in space and time to another place in space and time. For example, you write a secret message on a rock and throw it to a friend. But energy and information can also be delivered without the actual movement of bulk matter from one space-time point to another—carried in the form of *waves*. For example, sound waves that carry energy and information can *propagate* from your mouth to your friend's ear through the air. An electromagnetic radio wave can propagate energy and information from a vehicle on Mars to Earth through the vacuum of space. The wave method of transporting energy and information involves producing a physical disturbance in a continuous medium (or an electromagnetic disturbance in a vacuum), with the *disturbance* moving out through the medium (or vacuum) from one space-time point to another. In this case, no bulk matter actually moves through space-time. What moves is the disturbance—a wave propagates.

Many familiar types of waves require a medium to propagate. Large ocean waves caused by wind action on the surface (and sometimes by rock slides or seafloor earthquakes) propagate on the water surface and break on the shore. Smaller water waves can be propagated on the surface of a lake, initiated by a falling raindrop, a thrown rock, a bobbing duck, or a speeding boat. Sound waves can propagate in solids. For example, with your ear against a railroad track, you can hear an approaching train. Sound waves can propagate in a liquid. Whale sounds can travel for miles in the ocean. Sound waves can travel in gases. Musical sounds travel from the orchestra to your ear through the air in the concert hall. Waves can be made in a guitar or violin string by plucking or stroking. The string vibrations are sent through a bridge to the guitar or violin body, and the vibrating body produces sound waves in the air that propagate to your ear as sound. You can make waves in a stretched rope or Slinky by shaking one end. You can make waves on a drumhead by beating on the drum with a stick. Waves can be made in a metal or wood rod (xylophone) by hitting the rod with a small hammer. Seismic waves produced by an earthquake can travel both through the interior and on the surface of the earth. You can make waves in Jell-O by shaking the bowl. *Waves such as these that require a medium to propagate (and to exist) are called mechanical waves.*

But there are also waves that do not require a medium to propagate. Called *electromagnetic waves*, these waves can propagate through empty space—through nothing. An electromagnetic wave does not require a medium because it is made up of oscillating electric and magnetic fields that can self-sustain each other and move in a wavelike way. One source of electromagnetic waves is *accelerating or decelerating electric charge,* like in a radio broadcasting tower or power line. But most electromagnetic waves are produced by atoms and molecules during the process of making quantum transitions from higher to lower energy states. Electromagnetic waves come in an incredibly wide spectrum of categories. There are *visible electromagnetic waves* (visible light) associated with all the colors of the rainbow. These electromagnetic waves are detectable by the human eye. But there are many other categories of electromagnetic waves that the human eye cannot detect. These *nonvisible electromagnetic waves* (nonvisible light) include ultraviolet and infrared, radio waves, X-rays, microwaves, and gamma rays, to name only a few.

Both mechanical and electromagnetic waves can propagate in different spatial dimensions. In fact, all waves can be categorized as one-, two-, or three-dimensional. One-dimensional waves are waves that travel along a line without propagating out in two or three dimensions. Good examples of one-dimensional mechanical waves include waves propagating in a stretched rope or Slinky or sound waves confined to move along a metal rod. A good example of a one-dimensional electromagnetic wave is a laser light beam. Two-dimensional waves are waves that are confined to propagate on a two-dimensional surface or within a two-dimensional space. A ringlet of water waves propagating on the surface of a lake and waves moving across a drumhead are good examples of two-dimensional mechanical waves. Microwaves confined to a two-dimensional metal waveguide are examples of two-dimensional electromagnetic waves. Most mechanical and electromagnetic waves propagate in three dimensions. Sound waves propagating in air are three-dimensional mechanical waves. Visible light coming from a lightbulb and reflecting off objects in a room is propagating in three dimensions. Radio waves received by your car antennae or the electromagnetic waves that you use to cook food inside a microwave oven are also good examples of three-dimensional electromagnetic waves.

Unlike a baseball being thrown from one person to another, no bulk matter actually moves through space in wave motion. This is the single most important difference between particle motion (baseballs) and wave motion. So what moves in a wave? In a mechanical wave, it is a disturbance in the medium that moves. Once a disturbance is created at a given place in the medium, the adjacent part of the medium, being directly connected to the part of the medium where the disturbance was made, starts moving. This adjacent part does the same thing to its neighboring parts, and this *chain reaction* (the wave) propagates. This is assuming, of course, that there are no breaks in the medium. The movement of this disturbance through the medium is the wave. If the source of the disturbance is short-lived (say, a single pebble dropped into a lake), the spatial extent of the wave as it propagates through the medium will be small. Such a well-localized wave is often called a *wave pulse*. On the other hand, if the source of the disturbance continues to oscillate the medium for a number of cycles (say you shake the end of a Slinky back and forth a few times), the spatial extent of the wave as it propagates through the medium will be longer and will "show" a few cycles. Such a finite but extended wave is often called a *wavetrain*. It is also convenient and important to consider a third situation, in which the medium is being oscillated regularly and continuously (in theory) or at least for a large number of repeated cycles (in practice). In this case, the resulting wave as it propagates through the medium will have a very large spread and show many repeated cycles. Such a wave is called a *periodic wave*. These three categories (pulse, wavetrain, and periodic) apply to electromagnetic waves as well, but in this case it is not the duration or oscillatory nature of the medium disturbance that counts (since no medium is required), but the time extent and oscillatory nature of the source of electromagnetic waves (accelerating and decelerating electric charges or atomic and molecular transitions).

While these three categories lose some precision, especially in cases of overlap, they are very convenient in practice. Slapping the table with your hand produces a two-dimensional wave pulse in the table and a three-dimensional wave pulse of sound in the air. Voicing a vowel sound or musical note will produce a three-dimensional, periodic sound wave in the air. A laser beam is a one-dimensional, periodic electromagnetic wave.

Waves are also described in terms of how the movement of the medium relates to the direction of propagation of the wave. If the medium moves up and down *perpendicular* to the direction that the wave disturbance moves through the medium, the wave is said to be a *transverse wave*. A wave made in a stretched rope or guitar string is a transverse wave. By the way, all electromagnetic waves (visible and nonvisible light) are transverse waves, not because of any up-and-down oscillations of a medium, but because the electric and magnetic fields themselves oscillate perpendicular to the direction of propagation. If the medium moves back and forth along the *same direction as* the propagation direction, the wave is said to be a *longitudinal* wave. A longitudinal wave can be made to propagate in a Slinky by moving the end of the Slinky in and out along the direction of the Slinky. Actually, a Slinky can be made to carry transverse waves as well by shaking the end up and down perpendicular to the line of the Slinky. This is not true for a wave in a stretched rope—only transverse waves propagate. As they say, you can't push on a rope. Sound waves in air are longitudinal waves, where the air molecules move back and forth in the direction of motion of the wave. Indeed, gases can propagate only longitudinal waves. This is also true of sound waves propagating through liquids—only longitudinal waves are possible. Surface water waves are rather unique—the motion of the water is a combination of up-and-down (transverse) and back-and-forth (longitudinal) motion. Solids can also propagate both types of waves. Actually, solids can also propagate a third type of wave—called a *torsional* wave. In a torsional wave, the medium *twists* back and forth in a plane perpendicular to the direction of propagation. To imagine this type of wave, consider a stretched Slinky again. Instead of moving the end of the Slinky up and down to produce a transverse wave or moving the end back and forth to produce a longitudinal wave, you now grab the end and twist it clockwise and counterclockwise. This motion will propagate a twisting disturbance down the Slinky.

So how fast do mechanical waves move through a given medium? Part of the answer must come from considering how effectively the disturbance is transmitted from one part of the medium to its adjacent parts. In other words, the speed of propagation must be directly related to how well the medium is connected to itself. If adjacent parts of the medium are close together and rigidly connected, as in a solid, you would imagine that such a medium would propagate faster wave speeds. On the other hand, a sound wave in a gas in which the molecules are far apart and only loosely connected would propagate slower waves. And a sound wave in a liquid would be somewhere in the middle. This is indeed correct. Generally, longitudinal sound waves travel fastest in solids, not as fast in a liquid, and slowest in gases. For example, the speed of sound waves in aluminum is about 6420 m/s, in liquid water about 1497 m/s, and in room-temperature air about 340 m/s. As an aside, it is interesting to note that transverse and longitudinal waves actually travel at slightly different speeds in the same solid (since the atoms interact differently when compressed together than when sheared over each other).

The other part of the answer is based on the inertia factor. As discussed in some detail Chapter 1, objects with more mass are more stubborn, resulting in smaller accelerations for the same force. This would mean that all else being equal, denser media should propagate slower wave speeds. Indeed, under the same temperature and pressure conditions, the speed of sound in carbon dioxide gas versus air (nitrogen and oxygen) versus helium gas shows this dependency. The speed of sound is slowest in the densest gas (carbon dioxide), fastest in the least dense gas (helium), and somewhere in the middle for air. *In general, the speed of a mechanical wave depends directly on the strength of the binding (interactions at the atomic level) within the medium and inversely with the density of the medium.*

So how fast do electromagnetic waves travel through empty space? The answer is simple. *All electromagnetic waves (visible and nonvisible) travel at exactly the same speed in a vacuum—namely at the speed of light—approximately 300 million meters per second or 186,000 miles per second.* The exact value is 299,792,458 meters per second. Of course, it is not the medium here that is determining the propagating speed (since no medium is required), but the internal dynamics of the self-sustaining electric and magnetic fields that produce this constant speed. When electromagnetic waves travel through a medium (say, light propagating through glass), the oscillating electric and magnetic fields interact with the charges in the medium, and the speed is *reduced* from this vacuum value.

To summarize, both wave motion and particle motion carry energy and information—but they do so in very distinct ways. In wave motion, a physical disturbance in a continuous medium (a mechanical wave) or an electric-magnetic disturbance in empty space (an electromagnetic wave) propagates in space-time, while in particle motion, matter physically moves in space-time. Waves are everywhere. Familiar mechanical waves include surface water waves, seismic waves, Jell-O waves, rope and Slinky waves, as well as sound waves in solids, liquids, and gases. Familiar electromagnetic waves include all the colors of the visible light spectrum as well as a much larger spectrum of nonvisible light, including infrared, ultraviolet, radio, microwaves, X-rays, and gamma rays. Both mechanical and electromagnetic waves are often characterized by the dimension of the space in which they propagate (one, two, or three dimensions), by the characteristic motions of the medium or fields (transverse, longitudinal, or torsional), and by the extent and oscillatory nature of their waveforms (pulse, wavetrain, or periodic). The speed of a mechanical wave propagating in a given medium depends only on the nature of the medium—increasing with stronger internal binding of the atomic constituents and decreasing with media density. The speed of all electromagnetic waves in a vacuum is fixed at the constant speed of light.

Periodic Waves: Wavelength, Frequency, and Amplitude

Many of the most ubiquitous and interesting waves in nature—including most sound and light waves—are periodic. Such waves are created by a source that oscillates the medium (or oscillates electric charges within a vacuum) in a continuous and repetitive way. A "photograph" taken of a propagating periodic wave would show an extended waveform made up of repeating spatial units. The unit is called "one cycle" of the periodic wave. For example, consider a periodic wave that you have created in a stretched rope by shaking the end of the rope up and then down, up and then down, over and over again. The repeating unit (the cycle) might be taken to be the segment of the waveform from one crest of the wave to the next crest. In general, one cycle can be defined between any two repeated points in the periodic waveform. The spatial length of this cycle (measured, say, in meters) is called the *wavelength* of the periodic wave.

As the periodic waves propagates through the medium (or space), a given number of these cycles will pass a given point in the medium (or space) per second. The number of cycles that pass a given point per second is called the *frequency* of the periodic wave. Frequency is measured in units of cycles per second, a unit called a hertz (Hz for short), named after Heinrich Hertz (1857–1894). For example, say you are standing in a lake and watching a periodic water wave move by you and you notice that 3 cycles of the wave are hitting your leg per second; the frequency of the wave would be 3 cycles per second or 3 cycles/s or 3 Hz.

It is also important to realize that the frequency of a periodic wave is identical to the oscillating frequency of the source. This means that if the source of the periodic waves is shaking the medium a given number of times per second (the frequency of the oscillating source), then the frequency of the periodic wave produced will be the same.

The wavelength gives a measure of the spatial extent of one cycle of a periodic wave, but we also need a parameter to help us both describe and quantify the *height* of the cycle. The convention is to define the height of the wave, usually called the *amplitude* of the wave, as the maximum distance the medium gets displaced from its equilibrium position (the position of the medium when no wave is present). For electromagnetic waves, the maximum strength of the oscillating electric field is used.

As two- or three-dimensional waves move away from their source, the amplitude of the waves *decreases* with increasing distance from the source. This is a direct result of the fact that the original energy that went into producing the wave at the source site becomes more and more spread out as the wave propagates out in two or three dimensions. Drop a pebble in a lake and watch the two-dimensional ring of wavelets move out across the surface. As the waves move farther away from the source (the pebble location), the height of the waves becomes smaller and smaller, and eventually so small that the waves disappear. For three-dimensional sound waves propagating in air or for light propagating in a vacuum, this leads to the common observation that the intensity of the sound (loudness) or the intensity of the light (brightness) decreases with distance from the source. This is true because the loudness of sound or the brightness of light is related directly to the amplitude of the wave. Interestingly, this decrease in amplitude does not happen for one-dimensional waves (say, sound propagating down a railroad track or a laser beam crisscrossing a room). One-dimensional waves show *no decrease* in amplitude with distance from the source. This is a result of the fact that the original energy that went into producing the wave at the source has no chance to spread out as the wave propagates. In reality, even in one-dimensional waves, energy is dissipated for other reasons (friction, heat, etc.) and even one-dimensional waves will lose some intensity (but not nearly as rapidly as multidimensional waves) with distance from the source.

A very important relationship exists among the frequency, wavelength, and speed of a periodic wave. To determine this relationship, imagine producing a mechanical periodic wave with an oscillating source of low frequency, say, 2 oscillations per second, in a medium that propagates waves at a speed of 20 meters per second. A snapshot of this wave would show exactly 2 full cycles of this wave for every 20 meters of distance. Consequently, the length of 1 cycle, the wavelength, would be 10 meters long. What happens when the frequency is increased to a higher frequency, say 4 oscillations per second. Since the speed of propagation does not change, this means that now there will be 4 full cycles of the wave for every 20 meters of distance. In other words, the wavelength will decrease to 5 meters. Notice that by doubling the frequency (from 2 to 4 oscillations per second) the wavelength is halved (from 10 to 5 meters). Indeed, there is an *inverse* relationship between the frequency and wavelength in a given medium. *As the frequency of a periodic wave increases, the wavelength will show a corresponding decrease.* Of course, the reverse is also true. Decreasing the frequency will increase the wavelength. This inverse relationship between frequency and wavelength can be expressed mathematically:

$$(\text{frequency}) \times (\text{wavelength}) = (\text{wave speed})$$

This relationship is also valid for electromagnetic waves in a vacuum, with the wave speed set at 300,000,000 meters per second.

Wave Reflection and Transmission

A mechanical wave disturbance depends on a continuous and well-connected medium to maintain its propagation. But what happens to the propagating wave when it comes to the edge of its medium or strikes the boundary between two different media? For example, what happens to a sound wave in air when it strikes a wall? What happens to an ocean water wave when it strikes a breakwater? What happens to the wave that you have made at one end of a stretched rope or Slinky when it comes to the other end held by your friend? What happens to a wave that is generated in a drumhead or guitar string when it comes to the rim of the drum or to the end of the string? In most cases, the wave performs a very interesting division. Some part of the wave gets reflected and propagated back into the original medium, and the remaining part gets transmitted and propagated forward into the next medium. How much of each depends on the specific details of how well the two media are connected as well as the level of difference in the wave propagation properties of the two media. For example, sound waves are nearly fully reflected from a marble wall, but some sound is both reflected and transmitted when striking a wood wall. Light waves show nearly full reflection off a mirror surface but are both reflected and transmitted when striking a piece of glass. This reflection and transmission property is unique to wave motion and is another indicator that separates particle motion (baseballs) from wave motion. Baseballs can be either reflected or transmitted between two media, but they cannot do both at the same time.

Superposition Principle

When two or more pieces of matter (say, two football players) run into each other, they collide and often bounce off each other. They impede each other's motion. Waves, on the other hand, do not collide or bounce off each other, they simply move through each other unimpeded. This can be easily demonstrated by dropping two pebbles into a lake at the same time a short distance from each other. As you observe the two sets of waves moving out from each of their centers, you will notice that the expanding ringlets, even when they cross each other, continue their circular propagation. There is also evidence all around us for the simple but unique behavior of waves. Think about a dinner conversation (sound waves) involving two couples sitting across from each other at a table. The couples can carry on separate conversations without any difficulty. The two sets of sound waves just propagate through each other without any kind of collision or interference. If they didn't, the conversations would be greatly distorted. The same is true of electromagnetic waves. For example, imagine what the world would look like if light waves somehow collided and bounced off each other. Light propagating to your eyes from various objects in the room would be greatly distorted and you would not recognize the objects. The fact that waves do not interfere with each other's motion and can pass through each other unimpeded is another important property that separates particle motion (baseballs) from wave motion.

Related to this phenomenon, multiple waves have a unique way of occupying the same place in space and time. Indeed, what happens when two or more waves are propagating in the same medium (or through a vacuum) and are overlapping each other? What does the medium look like in this overlap region at a particular time? To imagine the resultant, picture two, identical, single-hump, transverse wave pulses moving toward each other in opposite directions along a stretched rope. We know that they will pass through each other, but what happens when they are fully overlapping, when they both meet for a moment in the middle of the rope? You observe a wave pulse of *double* amplitude (the waves *add*) during the moment they pass through each other. If one of the two wave pulses happens to be a *downward* pulse, with its displacement *below* the equilibrium position of the medium, then when the two waves overlap, the waves will momentarily give a resultant with no amplitude at all. In other words, the waves add again, but the positive (+) amplitude of one when added to the negative (−) amplitude of the other gives zero amplitude. If the waves are only partially overlapping, the net effect is determined by adding the contribution of each wave at each point in the medium. *In general, when two or more waves are propagating in the same medium (or through a vacuum), the net effect is simply the sum of the waves.* This important wave property is known as *the Superposition Principle.*

Activities

Chalkboard Waves: You will need a chalkboard (or a whiteboard), a piece of chalk (or a marking pen), and a chair with rollers. Have one student sit in the chair directly adjacent to the chalkboard. With chalk in hand, this student is to move the chalk up and down on the chalkboard while another student pushes the chair at a steady rate along the chalkboard. Observe the periodic waveform that is left on the chalkboard. Identify the amplitude and wavelength. Now have the student in the chair move the chalk up and down at a faster rate (higher frequency) with the same pushing speed. Note the smaller wavelength for the higher frequency. Try slower rates and note the larger wavelength. Try making waves of larger and smaller amplitudes. Can you make more complex periodic waves? Can you make a wavetrain? Can you make a wave pulse?

Drawing Waves on a Roll of Paper: You will need to locate an index card, some masking tape, a pen or pencil, and a smooth desktop. You will also need to purchase a roll of paper (the narrow kind used in cash registers). You first need to make a "sleeve" out of the index card through which you can pull the paper coming off the roll. To do this, fold opposite edges of the index card over and under the sides of the paper, giving a little extra room so that the paper can be pulled through easily. Tape this sleeve to the tabletop. Cut a one-inch slit in the middle of the index card (the sleeve) just wide enough to accept the tip of the pencil or pen. The slit orientation must be perpendicular to the direction the paper will be pulled through the sleeve. Place the roll of paper on a smooth desktop and lay a length of paper out on the table.

Thread the paper through the sleeve. Have one student move the pencil back and forth inside the slit as another student pulls the paper at a steady rate through the sleeve. Observe the periodic waveform that is left on the paper. Identify the amplitude and wavelength. Now have the student move the pencil up and down at a faster rate (higher frequency) with the same pulling speed. Note the smaller wavelength for the higher frequency. Try slower rates and note the larger wavelength. Try making waves of larger and smaller amplitudes (cut different slit lengths). Can you make a wavetrain? How about a wave pulse? Can you make more complex periodic waves?

Student Waves: You will need to find a space where all your students can stand in a circle. Have your students hold hands in a big circle, facing the middle. Designate one student as the source of the wave. This person starts the wave by moving one of her arms up and then down. When the student holding hands with this source person feels his arm being raised and lowered, he raises and lowers his other arm, and this cycle is repeated with the next person, and so on. This *transverse wave pulse* can continue around the circle as many times as desired. Try to model a wave with smaller amplitude (students do not raise their arms as high). After some practice, see if students can make the pulse move around the circle at a faster propagation speed, at a slower propagation speed. Make a break in the circle to see if the students can model wave reflection. When the wave arrives at the last person (the break), have that person start the wave again in the opposite direction. See if the class can perform multiple reflections back and forth through the student medium. You can also model a *longitudinal wave pulse* by having students stand in a circle front-to-back with the hands of one student on the shoulders of the next. Have the source person initiate the wave by pushing and pulling gently on the shoulders of the student in front of her. The source person should also take a step forward and backward as she pushes and pulls on her partner's shoulders. This pushing-pulling and forward-backward stepping is repeated around the circle. Try changing the propagation speeds. Try reflections. Try having your students model a wave with smaller amplitude by taking a shorter step forward and backward. You can also attempt to model *reflection and transmission* by having students make a single line with the girls on one end of the line and the boys on the other. With everyone holding hands (including the boy and girl in the middle), the teacher initiates the wave at one end of the line (say at the boys' end). As the wave propagates through the boy medium and arrives at the junction with the girls, the junction girl propagates the wave forward through the girls (transmission) but simultaneously reflects the wave back though the boys (reflection).

Giant Slinky and Slinky Rope Waves: You will need to buy a giant Slinky (Super Slinky, Arbor Scientific) and a Slinky rope (Snaky, Arbor Scientific) from a science supply house. You will also need a smooth floor. The giant Slinky and the Slinky rope provide wonderful media for making waves and observing a variety of wave properties. Stretch the giant Slinky (up to 20 feet) along the floor with a student holding each end. Make transverse waves by having one student shake her end of the Slinky back and forth along the floor perpendicular to the Slinky. Make longitudinal waves by having one student shake his end in and out along the Slinky. Try making wave pulses. Try making periodic waves. Try different amplitudes. Try varying the frequency of the periodic wave and noting the corresponding change in wavelength. Observe the wave reflection at the ends of the Slinky. Increase the tension in the Slinky or Slinky rope by stretching the rope even more (this also decreases the mass per unit length of the rope) and watch the wave speed increase. Have the students at each end send a wave pulse down the Slinky to see if the pulses propagate through each other and add together when overlapping. Try connecting the giant Slinky and Slinky rope together and propagating a wave pulse down the entire length. Observe what happens to the wave when it arrives at the connection point. Do you observe both transmission and reflection at the junction?

Waves in a Rubber Rope: You will need to purchase a 20-foot length of rubber tubing from a science supply house (Pure Gum Amber Rubber Tubing, Sargent-Welch). Tie one end of the rubber tubing to a doorknob or some other permanent fixture in your classroom. Holding the other end, stretch the tubing across the classroom. Make transverse waves by shaking the end of the rubber rope up and down perpendicular to the rope. Try making wave pulses. Try making period waves. Try varying the frequency of the periodic wave and noting the corresponding change in wavelength. Try different amplitudes. Observe the wave reflection at the fixed end. Increase the tension in the rope by stretching the rope even farther (this also decreases the mass per unit length of the rope) and watch the wave speed increase. Instead of fixing one end of the rubber rope to a doorknob, let two students stretch the rope between them. Have each send a wave pulse down the rope to see if the pulses propagate through each other and add together when overlapping.

Water Waves With an Overhead Projector: You will need to locate an overhead projector and screen. You will also need to purchase a shallow, clear, flat-bottomed water tray. The clear plastic picture frame boxes used to mount photographs work very well. The tray should be large enough to cover the overhead projector light plate. You will also need to locate some clay (plasticine works best), two pencils, and two plastic rulers. Place the water tray on the overhead projector and fill it with about one-quarter inch of water. Turn on the overhead and focus the projector until you can see waves in the tray focused on the screen. Produce some circular wave pulses by dabbing a pencil tip into the water; watch the screen. Try making a periodic wave by dabbing the pencil in and out of the water a few times. Change the dabbing rate (frequency) and observe the change in wavelength. Observe the circular waves reflect off the side of the tray. Are the reflected waves circular also? Try making two sets of circular waves by dabbing two pencils into the water a short distance from each other. Do you observe the two sets of waves passing through each other? Instead of using pencil points to make the water waves, try dabbing the edge of a ruler in the water and making "flat" waves. Observe how these waves propagate and reflect. Make two sets of flat waves at an angle to each other and observe them as they cross each other. Do they propagate through each other? Try reflecting the flat waves off the side of the tray. Are the reflected waves flat also? Use the clay to make various shaped barriers in the water (e.g., C-shaped and flat). Send both circular (use the pencil) and flat (use the ruler) waves at these barriers and observe their reflections. Send some flat waves toward a small circular barrier (say, a 1-inch-wide ball of clay) to see if you can observe the waves *diffract* around the barrier. Make a small opening in an otherwise flat barrier and send some flat waves through the opening. Can you see the waves diffract as they propagate through the opening?

Tuning Fork Play: You will need to locate or purchase a set of tuning forks (Tuning Forks, Arbor Scientific). You will also need to locate a Ping-Pong ball (or any table-tennis ball) and a cup of water. Strike a tuning fork and listen to the sound waves being generated by its vibrating prongs. Students should bring the tuning fork closer to one ear to hear the sound get louder. Move the fork away from the ear and the sound gets softer. The amplitude of the sound waves (related to the loudness sensation) decreases with an increase in the distance from source to the receiver (ear). Students can feel the vibrations of the tuning fork with their fingers. Student can also touch a vibrating tuning fork to the surface of a glass of water and see the spray. Students can also touch a vibrating tuning fork to the water surface in a tray of water on an overhead projector (see "Water Waves With an Overhead Projector," above) and observe the water waves produced. Students can bring a Ping-Pong ball, dangling from and taped to a piece of string, up to a vibrating tuning fork and watch it bound off. If you have access to a stroboscope, you can "freeze" the vibrating prongs for observation by setting the strobe's frequency close to the tuning fork's frequency.

Students can press the handle of a vibrating tuning fork against a tabletop or other object to hear a louder, amplified sound (although the sound will last for a shorter period of time). Students can press the handle of a vibrating tuning fork to their chin or the bone just behind their ear and "hear" the sound that propagates through the skull directly to the middle ear.

Garden Hose and Stethoscope: You will need to locate a garden hose and borrow a stethoscope from your doctor. Since the hose and the stethoscope channel the sound waves in one-dimensional tubes, the amplitude of the waves (related to the loudness sensation) will not decrease very much with distance (compared to three-dimensional waves, whose amplitude decreases rapidly with distance). This allows you to talk over larger distances with a hose or listen to softer sounds with the stethoscope. One student whispers into one end of the hose while a classmate listens at the other end. Run the hose from one room to another and have students communicate through the hose. Have students listen to their own or a classmate's heart beat through the stethoscope. Students should try to listen to other soft sound sources (ticking watch, etc.) through the stethoscope as well.

Tubes and the Ticking Watch: You will need to locate a stopwatch (or any watch that produces a ticking sound), some cardboard mailing tubes, and duct tape. Place the watch on the floor and place one end of a mailing tube over the watch. You should be able to hear the ticking through the mailing tube. Connect two or more tubes together with duct tape and listen to the watch again. Since the tubes channel the sound waves into a one-dimensional space, the amplitude of the waves (related to the loudness sensation) will not decrease very much with distance (compared to three-dimensional waves whose amplitude decreases rapidly with distance). This allows you to listen to softer sounds over larger distances. On a related note, students might try to connect a series of tubes together across the classroom (or along a bulletin board) and see if they can talk through it with each other.

Coat Hanger, Spoons, and Grill Chimes: You will need to locate some string, a metal coat hanger, metal spoons, and a barbecue grill (not the whole barbecue, just the metal grill part). Tie the hook part of the metal coat hanger to the middle of a three-foot-long piece of string. Wrap a few turns of one end of the string around the pointing finger of one hand and wrap a few turns of the other end of the string around the pointing finger of your other hand. Insert the pointing fingers in each ear as you dangle the coat hanger below you. Have the dangling coat hanger strike a nearby object (chair, table, etc.) and listen to the sound waves that travel through the hanger, through the string, and into your ears. Repeat using dangling spoons or other metal objects. Try hanging the grill from the string and repeat, but this time have someone stroke across the grill with a spoon. Similar to the last two activities, the vibrating coat hanger, spoons, or grill produces one-dimensional waves that propagate through the strings directly to your ears.

Everyday Examples

Pseudo Waves: There are a number of phenomena that *simulate* wave motion, such as "the wave" at a sporting event moving around the stadium, wind blowing "waves of grain" over an open field, or a "domino wave" caused by a chain of dominoes falling successively into each other. Since these happenings are disturbances that are not propagating because of a continuous and connected medium, they are not mechanical waves. But these pseudo or simulated waves are interesting to explore.

Ocean Waves: Large ocean waves are initiated by wind blowing over the ocean surface. As an ocean wave approaches a shoreline, the front part of the wave experiences the shallow water before the back part of the wave. Since water waves move slower in shallower water, the front part of a wave is slowed relative to the back part. This causes the wave to "break." For the same reason, shallow reefs also cause ocean waves to break over them. Besides wind-generated waves, water waves can be produced by earthquakes and rock slides.

Tsunamis: These are extremely powerful ocean waves caused by earthquakes in the ocean floor.

Waves in a Swimming Pool: People playing in or jumping into a swimming pool create lots of surface water waves. Many pools (especially those used for swimming and diving events) are designed with a special gutter around the edge to minimize the reflections of these waves. There are a number of water parks and some aquariums that mechanically generated waves in a pool to simulate ocean waves.

Waves in a Bathtub: You can create large waves in a bathtub by sloshing back and forth in the tub.

Raindrops and Pebbles in a Smooth Lake: Raindrops falling or pebbles thrown into a smooth lake create beautiful ringlets of surface water waves.

Rising Trout: Fly fishermen and fly fisherwomen often look for small circular waves on a lake or in a river as a sign that a trout (or another type of fish) has come up from below and taken a fly off the surface. They sometimes refer to this event as the "ring of the rise."

Boats and Swimming Ducks: Motorboats and swimming ducks produce V-shaped water waves in their wake. Large seagoing vessels can produce very large waves in their wake.

Soap Bubble Film: You can make very slow moving waves in soap bubble films.

Waves in a Suspension Bridge: There have been a number of cases where very large torsional waves—initiated by the wind—have occurred in suspension bridges. The Tacoma Narrows Bridge in Washington state was destroyed by such waves.

Waves in Towels and Sheets: When you shake the end of a towel or bedsheet, waves propagate through the material.

Waves in Telephone Lines: On a windy day, you can sometimes see waves propagating in telephone lines.

Passing Trains and Trucks: When a train or heavy truck passes close to you, you can often feel the waves propagating through the earth to your feet. In some cases, these waves can cause your house to shake.

Water Waves and Barriers: When water waves coming from the ocean propagate through an opening in a harbor breakwater, they diffract (bend in a circular pattern) as they enter the harbor. Ocean waves can also be seen diffracting around the end of a breakwater or tip of an island or peninsula. Ocean waves also bend (diffract) around small islands.

Seashore Superposition: Sometimes when a wave breaks on the shore, part of the wave gets reflected back into the water. You can see this wave pulse moving away from shore. When this reflected wave meets up with an incoming wave, the moment of overlap can create a large wave (the superposition principle).

Seismic Waves: Seismic waves (earthquake waves) are categorized into two families: *body waves* and *surface waves.* There are two types of body waves: *Primary waves* (P-waves) are longitudinal waves, consisting of alternating regions of compression and expansion acting in the direction of the wave travel. They propagate through all parts of the earth (solid and liquid). In the uppermost portion of the earth's crust, P-waves travel at around 6 km/s and are the first waves to arrive at a seismic station after an earthquake. *Secondary waves* (S-waves) are transverse waves that propagate through the solid parts of the earth, but not through the liquid outer core. Liquids and gases cannot propagate transverse waves. In the uppermost portion of the earth's crust, S-waves travel at around 3.5 km/s and arrive after the P-waves. *Surface waves* are seismic waves that propagate on and are restricted to the surface of the earth. They travel slower than P- and S-waves and are the last to arrive at a seismic station after an earthquake. The reflection and refraction of P- and S-waves are used to determine the structure of the interior of the earth. Surface waves can be used to probe the structure of the crust.

Dead Spots in a Microwave: There are some areas in a microwave oven that are less efficient in cooking your food. This is why most microwave ovens have a rotating platform or why a recipe's instructions for microwave cooking suggest that you rotate the food periodically. There are places in the microwave oven where two or more microwaves are superimposing and the resulting wave effect (due to the addition of the waves) is small.

Coffee Waves: Waves are sometimes produced in a cup of coffee when you place the cup firmly down onto a table.

Electromagnetic Waves: Visible light, infrared, ultraviolet, radio, microwaves, and X-rays are used in many technological applications. See Chapter 4, "Light and Electromagnetic Waves," for specific examples.

SOUND WAVES

Concepts

Sound waves can propagate as longitudinal waves in solids, liquids, and gases. While solids can also propagate transverse sound waves, liquids and gases cannot. Longitudinal periodic sound waves are created when the source of the sound (vibrating tuning fork, stereo speaker, vocal cords, etc.) periodically compresses and expands the medium. It is the propagation of these compression and expansion regions that is the sound waves.

Here are two helpful ways to visualize this longitudinal propagation. Imagine a long tube filled with air with a plunger at one end. As you push the plunger in and then out of the end of the tube, you will compress the air on the inward stroke and expand the air on the outward stroke. These compressions and expansions propagate down the tube. For sound waves in air, the compression regions have a slightly larger gas pressure than the ambient pressure (air pressure in the tube without the wave), while the expansion regions have slightly *smaller* pressure than ambient pressure. As a result, it is often convenient to think of a sound wave as a propagating wave of larger and smaller pressure variations. Indeed, it is these pressure variations of the sound wave in air as they impact the eardrum that set the eardrum into vibration to initiate the hearing process.

Another way to visualize a longitudinal wave is by using a large Slinky stretched between two students across a smooth floor (see "Giant Slinky and Slinky Rope Waves," above). One student periodically compresses and expands one end of the Slinky to produce a wave in the Slinky of alternating regions of compression (Slinky coils compressed closer together) and expansion (Slinky coils stretched farther apart).

Humans can detect sounds waves with frequencies from around 20 cycles per second (20 Hz) up to around 20,000 cycles per second (20,000 Hz). The lower frequencies are heard as lower pitch sensations and the higher frequencies are heard as higher pitch sensations. Indeed, the ear is a rather amazing device. Your eardrum can be set into vibration by a 20,000 Hz tone, which also means your eardrum will be vibrating 20,000 times per second! And you can detect this as a very high pitch sensation. The ear is most sensitive around 3000 Hz. The *audible range* in frequency (20 Hz to 20,000 Hz) does vary from person to person and definitely changes with age. As they say, by the time you are old enough to afford a good stereo system, you don't need one. The range from 0 Hz to 20 Hz (called the *infrasonic* range) and from 20,000 Hz and up (called the *ultrasonic* range) are not detectable by humans as a sound sensation, but these ranges do affect humans in other ways (see "Everyday Examples"). Of course, other animals use, produce, and detect sound waves in other frequency ranges (see "Everyday Examples").

The audible frequency range for humans, 20 Hz–20,000 Hz, can also be represented in terms of a range in wavelength. Since (frequency) × (wavelength) = (wave speed), we can determine the range in wavelength by dividing the frequency into the wave speed. With the speed of sound in air (wave speed) at 340 meters per second, a 20 Hz sound wave would have the longest wavelength: 340/20 or 17 meters. A 20,000 Hz sound wave would have the shortest wavelength: 340/20,000 or about 1.7 centimeters. So the audible range in wavelength is from 1.7 centimeters to 17 meters.

The amplitude of a sound wave corresponds to the extent to which the medium gets compressed and expanded. For a sound wave in air, the strength of the maximum and minimum pressure variations (maximum in the compression region and minimum in the expansion region) is often used as a measure of the amplitude of the sound wave. While the frequency of a sound wave corresponds to the pitch sensation, the amplitude of a sound wave corresponds to the *loudness* (also referred to as *intensity* or *volume*) of the sound. The human ear can sense pressure variations as small as 1 part in 10,000,000,000 of atmospheric pressure (called the threshold of hearing or 0 decibels) up to about 1 part in 10,000 (1,000,000 times greater), where the loudness sensation becomes painful (120 decibels). Of course, for two- and three-dimensional sound waves, the amplitude, and consequently the loudness of the sound, decreases with increasing distance from the source.

Two waves can have the same pitch (frequency) and loudness (amplitude) but can vary greatly in their *quality.* Indeed, the sound of a clarinet and a violin both playing the same note at the same loudness level are still distinctly different sounds. They differ in the specific details of the pattern within the complex wave. In other words, while two sounds might have the same size repeating cycle (same wavelength and frequency), the details within that cycle can be different. And it is that detail that carries the quality of the sound, distinguishing one sound from another.

Activities

Dancing and Singing Toys: You will need to purchase two of these very amusing sound-activated toys from novelty stores. One of the two toys should be the singing bass (Billy Bass) and the other a toy that dances when activated by sound. Have students activate either of these toys by making sounds in the room. This activity provides an initial opportunity to discuss the production of sound (vocal cord vibrations), the propagation of the sound (waves in the air medium), and the detection of the sound by the toy. The toy has a receiver that converts the sound wave oscillations (pressure variations) to an electrical signal that is used to turn on a small battery-powered motor. Now comes the fun part. Without any other sounds in the room, have the sound waves from the singing bass activate the dancing toy!

Investigating Sound Toys: You will need to go to a toy store and purchase a variety of toys that are designed to produce sounds. Have students investigate these sound toys and determine how the sound is actually produced.

Record Player Fun: You will need to locate an old turntable and record (music). You will also need to locate some stiff paper, masking tape, and a few sewing needles. Roll the paper into a cone and use the masking tape to hold it together. Stick a sewing needle through the tip of the cone about 1/2 inch from the tip end. The sharp end of the needle should stick out about 1/4 inch. Turn on the record player. Now hold the cone such that the needle tip is resting gently on the rotating record. Hold the cone tangent to the record. While it will be reduced in quality, you should be able to hear the music loud and clear. The groove in the record has ridges and valleys that vibrate the needle. The needle vibrates the cone, which in turn vibrates the air and produces the sound. What do your fingers feel as you hold the cone? Try doing it without the cone, with only a handheld needle.

Mystery Sounds: You will need to locate tin cans with plastic lids. Place a different object (marble, rubber ball, penny, key, paper clip, etc.) or objects (sand, popcorn kernels, water, etc.) in each can and seal with the plastic lid. See how well your students can guess what is in each can by listening to the sound each makes when rattled.

Tuning Fork Play: You will need to locate a cup of water, a piece of stiff paper, and a Ping-Pong ball. You will also need to locate or purchase a set of tuning forks (Tuning Forks, Arbor Scientific) and, if possible, a strobo-scope (Digital Strobe, Arbor Scientific).

  A. Have students pick up a tuning fork and strike one prong against a table. They can investigate the vibration by placing the prongs on the surface of the water in the cup. The vibrating prongs will spray the water, and students may be able to see some surface waves in the water.
  B. Students can also feel the vibrations of the prongs with their fingers or tongue, or by holding the vibrating fork against a piece of paper.
  C. In a darkened room, set the stroboscope frequency to the fre-quency indicated on a tuning fork and observe the vibrating prongs in strobe light. You will be able to freeze-frame the motion to "see" the vibrating prongs.
  D. Tape one end of a piece of string to the Ping-Pong ball and suspend the ball in the air. Touch the vibrating tuning fork to the ball and watch it fly.
  E. Bring the vibrating tuning fork close to your ear and listen. Use dif-ferent tuning forks and note the connection between the vibration rate (stamped on tuning fork) and the pitch sensation.
  F. Touch the end of the stem, not the prongs, of the vibrating tuning fork to the skull bone behind your ear and listen to the sound that

is transmitted through your skull directly to the middle ear. Try different tuning forks. Place the stem of a vibrating tuning fork on the bony part of your elbow with a finger (same arm) in your ear. Can you hear the sound?

G. Place the end of the stem of the vibrating tuning fork (not the prongs) firmly against a tabletop and listen to the "amplified" sound. The tuning fork vibrations are transmitted to the tabletop (acting as a sounding board), which, in turn, transmits the sound to the air. You can also listen to the tuning fork by placing your ear directly on the tabletop.

Sound Waves in Liquids: You will need to locate an aquarium or sink, a cooking timer (the windup type, not electrical), two spoons, and an airtight food storage bag. Fill the aquarium or sink with water. Submerge the two spoons in the water. Click the two spoons together and listen to the sound produced. The sound waves propagate through the water, into the air, and to your ear. Now try this. Set the cooking timer to one minute. Place the cooking timer in the plastic bag and seal. Place the timer and bag underwater and wait for the timer to go off. Can you hear the sound of the timer?

Sound Waves in Solids: You will need to locate a yardstick or meter stick and a watch or clock that has a distinctive ticking sound. To investigate sound traveling through a solid, place your ear to the end of the yardstick and hold the ticking watch or clock firmly against the other end. Can you hear the sound? Repeat, but this time use a desktop instead of the yardstick. Instead of using a ticking clock or watch as the sound source, try tapping or scraping the yardstick or desktop with your finger. You can also place your ear to a wall and have a friend tap the wall.

Cup and String Telephones: You will need to collect various types of cups (paper, plastic, foam, metal, etc.), various types of string (nylon twine, monofilament fishing line, kite string, yarn, metal wire, etc.), a metal coat hanger, scissors, and paper clips. To make your telephone, poke a hole in the bottom of two identical cups. Pull a string through the bottom of each cup and tie a paper clip onto the end of the string that is inside the cup so that the string doesn't fall out. When talking and listening, make sure the string is taut so the sound can be transmitted easily. Investigate using paper cups, foam cups, plastic cups, and metal cups (all with kite string) to find out which cup makes the best telephone. Now try various strings (all with the same cups) to find out which type of string is the best. Here are some other investigations to try. (1) With a single set of telephones and both partners listening, have someone else pluck the string between them. Change the tension in the string and repeat. (2) Hang the metal coat hanger in the middle of the string. Have someone strike the hanger as the two partners both listen through their cups. (3) Loop the lines from two sets of cup telephones around each other. Now each person has a cup for talking and another for listening. (4) Make a party line by looping two or more sets together. Now many people can participate in the conversation. (5) Make a very long telephone system. How long can you make it and still transmit and hear the conversation?

Talking Tapes: You will need to purchase these interesting talking tapes from a science supply house (Talking Tapes, Science Is Fun, Educational Innovations). By rubbing a plastic tape that has been designed with a sequence of molded grooves, you can produce audible speech sounds in a paper cup attached to the tape.

Plucking a String: You will need to locate various types of string (nylon twine, monofilament fishing line, kite string, yarn, metal wire, rubber bands), a desk or chair, and scissors. Cut the strings into three- or four-foot lengths. Pick one of the string types and tie it securely to the desk or chair. Pull it taut with one hand and pluck it with the other. Only a very faint sound will be heard. The sound is not very effectively transmitted from the vibrating string to the air to your ears. Now take the free end of the string and wrap it around your index finger and place that finger in one of your ears. Puck the string again and listen. You will hear a significant difference in the sound intensity. The string vibrations are transmitted to your finger and directly to your ear. Try various types of string, different lengths of string, and different tensions. Shorter and skinnier strings (less mass) with more tension will vibrate at a faster rate, producing higher frequency sound (higher pitch sensation). Longer and fatter strings (more mass) with less tension will not vibrate as rapidly and will produce lower pitch sounds.

Pitch and Size: You will need to locate or purchase some of the following items: (1) nails made of the same material and thickness but of different lengths, string, and a spoon; (2) two metal rods (Singing Rods, Arbor Scientific) of the same material and thickness but of different lengths and a hammer; (3) a xylophone; (4) a guitar; (5) a set of toy wooden frogs (Wooden Percussion Frogs, Educational Innovations); (6) hanging chimes and a spoon; (7) a set of tuning forks (Tuning Forks, Arbor Scientific); (8) a set of bells; (9) six glass soda bottles and water; (10) sound tubes (Basic Boomwhacker Set, Educational Innovations); and (11) plastic straws and scissors. These 11 activities will all share a common theme. The pitch of the sound (the frequency of the sound) produced by various objects depends on the *size* of the object. *In general, smaller objects produce higher pitch (higher frequency) sounds than larger objects of the same material and shape.* The reason is this. When an object (such as a nail or guitar string) is struck, the sound waves propagate through the object at a set speed (the speed of sound for that medium) and reflect off the end. These reflected waves propagate back through the object and reflect again off the opposite end. These reflections continue back and forth in the object. The overall vibration of the object is a direct result of these multiple reflections. The waves in smaller objects have less distance to cover (at the same speed); consequently, the back-and-forth reflection rate (the vibration rate) will be higher than for larger objects. This means that the smaller object will produce a higher pitch sensation. This is also true of sound in air in a confined space (such as air trapped in a tube or straw). For example, smaller tubes will produce higher frequency/pitch sounds.

1.   Use the string to suspend three (or more) nails of different lengths from the edge of a table. Tie the string just below the head and suspend the nail vertically. Strike each nail with the spoon and listen to the different pitch sensations.

2.   Pick up one of the metal rods and balance it horizontally by holding it with your fingertips at the middle point. Strike one end of the rod with the hammer and listen to the pitch sensation of the sound produced. Repeat with the other rod.

3.   The xylophone has a series of metal bars of increasing length. Strike each in turn and notice the decreasing pitch as you move to longer and longer bars.

4.   Pluck one of the strings of the guitar and listen to the pitch sensation. Now shorten the string by pushing your finger down on the string and repeat. As the string gets shorter, the pitch increases.

5.   Make the toy wooden frogs croak according to the instructions. Notice the difference in pitch sensation between the smaller and larger frogs.

6.   The hanging chimes are just bars or tubes of different lengths. Hit each chime with the spoon and notice the difference in pitch sensation.

7.   Strike the tuning forks and notice that the higher frequency tuning forks are much shorter than the lower frequency forks.

8.   Ring the bells and notice that the higher frequencies come from the smaller bells.

9. Fill six identical glass soda bottles with different amounts of water. Blow gently over the top of each bottle opening to produce sound. Notice that the bottle filled with the most water (and hence with the smallest column of air) has the highest pitch and the bottle filled with the least water (and hence with the longest column of air) has the lowest pitch.

10. Follow the manufacturer's directions for the sound tubes. Notice that the longer tubes produce the lower pitches.

11. The first thing you need to do is make a reed for your straw. To do this, flatten about one inch at one end of the straw. With the scissors, cut a section off of each side of the flattened portion of the straw, making the end of the straw narrower than the main straw width. To play the straw reed, place the straw in your mouth and, with your teeth pressed lightly on the top and bottom of the reed just beyond the end of the cuts, blow. You will probably need to practice with different teeth pressure and placement to make it work. Once you can produce a steady sound with your straw, make additional straw instruments of different lengths. Notice that the longer straws produce the lower pitches. If you can locate straws of different cross-sectional sizes, you can use a larger or smaller straw as a trombone-like slide to continuously change the length.

Squealing Balloons and Boxes: You will need to locate a balloon and a box of cards. Blow up the balloon and try to make a squealing sound by releasing the air slowly through the highly constricted throat of the balloon by pulling outward on both sides of the throat. You will need to adjust the tension with which you are pulling to produce the squeal. This provides a good analogy for how the vocal cords produce sound. The air rushing through the vocal cords from the lungs is analogous to the air rushing out of the balloon through its constricted throat. The rushing air causes the vocal cords to vibrate and produce sound. Here the rushing air vibrates the throat of the balloon at a high rate, producing the squealing sound. Another way to produce a similar squealing sound and further reinforce the vocal cord analogy is to make a box of cards sing. To do this, open one end of the box and remove the cards. Place your mouth in the opening of the box and blow hard into it. The air will move through the box and rush out through the small opening between the box and the sealed flap. This rushing air will cause the sealed flap to vibrate at a high rate.

Wok Reflections: You will need to locate a large Chinese wok. In a quiet room, hold the wok in front of your face at arms length and talk into it. Move the wok toward your face as you talk to find a place where you hear the loudest sound. The concave side of the wok does a good job of reflecting and focusing the sound waves. Repeat with the convex side facing you. In this case, the sound is not focused and the reflected sound is not nearly as loud.

Stringed Instruments: You will need to locate a variety of stringed instruments (guitar, ukulele, banjo, violin, cello, etc.). Have students investigate how the pitch of the sound produced by vibrating string depends on the length, thickness (mass), and tension in the strings. To investigate the dependence on length, listen to the sound as you shorten the string (using a movable bridge or your finger). Shorter strings (thickness and tension being equal) will produce higher pitches. To investigate the dependence on thickness, locate an instrument that has different string thicknesses and adjust all the strings to approximately the same tension and length. Thicker (more massive) strings (length and tension being equal) give lower pitches. To investigate the dependence on tension, change the tension in a string to see how this changes the pitch. More tension (length and thickness being equal) gives higher pitches.

Sound Intensity: You will need to purchase a sound-level meter (Sound Level Meter, Sargent-Welch). Use the meter to measure the sound intensity (loudness) of different sounds in various locations (classroom, playground, busy street, band room, cafeteria, etc.). Also measure the sound level coming from sound sources whose loudness you can control (radio, boom box, TV set, etc.) for different volume settings and for different distances from the sources.

Microphone and Oscilloscope: You will need to borrow a microphone and an oscilloscope from your local high school physics teacher or college physics department. The oscilloscope allows you to "see" a picture of the sound waves detected by the microphone. Students can "see" their name as a sound wave and can make other sounds into the microphone to see the sound wave patterns for different sounds. This is also a wonderful way to explore the important properties of sound waves: (1) the frequency-wavelength dependence (higher pitches are shorter in wavelength) using tuning forks, musical instruments, vocal sounds, etc.), (2) the amplitude-loudness dependence (louder sounds have larger amplitudes) using sound sources of different loudness levels, and (3) the difference in the quality of different sounds that are similar in pitch and loudness using various musical instruments sounding the same note or by voicing different sounds (*ee* sound or *ah* sound) at the same pitch and loudness.

Sound Absorption: You will need to locate a battery-powered radio, boom box, or clock radio. You will also need to locate various items to act as sound absorbers (books, cardboard, bubble wrap, foam, clothes, blankets, pillows, insulation, etc.). Turn up the volume on the radio or boom box or clock radio and see how well you can muffle the sound by surrounding the radio with these different materials. Which materials work best?

Everyday Examples

Dead and Live Spots in Buildings: As a result of the reflections off the walls, floor, and ceiling, sound waves in a building can superimpose to create areas where the waves add to create a louder sound and other areas were the waves partly cancel to produce a softer sound.

Concert Hall Acoustics: The correct amount of *reverberation,* not too much and not too little, is important for quality concert hall performances. Reverberation is the multiple reflections of sound waves off all interior surfaces, including people in the audience. Acoustical engineers design the interior spaces with the correct amount of reflecting and absorbing materials in order to produce the optimum reverberation time.

Sound Waves in Bathtubs and Swimming Pools: With your head submerged in a bathtub or swimming pool, you can easily hear sounds. The next time you are in a bathtub or pool, submerge your head and rap on the side of the tub or pool with your hand.

Whales and Porpoises: Whales, porpoises, dolphins, and some other sea creatures communicate using sound waves in the ocean. For example, porpoises can produce sound waves from around 7000 to 120,000 Hz (compared to 85–1100 Hz for humans) and can hear sound waves in the 150–150,000 Hz range (compared to 20–20,000 Hz for humans).

Bats and Echolocation: Bats use echolocation (sound wave reflection) to locate their prey, primarily moths, and to negotiate their physical environment. Bats can produce sound waves in the 10,000–120,000 Hz range and detect sound waves from 1000 to 120,000 Hz.

Echoes: You can hear sound waves returning after reflecting off canyon walls and other large structures.

Sonar: Submarines use sonar (sound wave reflection) to detect other ships in the water. Scientists use sonar to map the ocean floor. Historians and treasure hunters use sonar to locate shipwrecks. Fishermen use sonar to locate schools of fish and underwater structures.

Ultrasound: Sound waves above the audible range in frequency are used in a variety of medical applications, including detecting heart problems, kidney stones, blood flow problems, and prostate cancer and imaging a fetus in the womb.

Infrasound: Sound waves below the audible range in frequency are all around us. Animals such as whales use this sound range for long-range communications. In some cases, infrasound can be "felt" by humans. Strong storms, ocean waves, airplane sounds, and winds in the mountains can generate infrasound that may cause sensations in humans.

Intercom Systems: Intercom systems in older homes were based on sound tubes running through the walls. Since one-dimensional sound waves do not lose much intensity with distance, people on different floors and in different rooms could communicate with each other by speaking and listening through these tubes.

Hitting a Nail: When you pound a nail into a piece of wood, the pitch of the sound increases after each strike of the nail with the hammer. As the nail shortens, the frequency of the nail vibration increases (see "Pitch and Size," above).

Animal Sounds: While humans generally produce sounds in the 85–1100 Hz frequency range and detect sounds in the 20–20,000 Hz range, other animals produce and detect sounds in different ranges. For example, dogs produce sounds from around 500 to 1100 Hz but can detect sounds over a much larger range, 15–50,000 Hz; frogs produce sounds from 50 to 8000 Hz and detect sound from around 50 to 10,000 Hz; and bats produce sounds in the 10,000–120,000 Hz range and detect sounds in the 1000–120,000 Hz range.

Male Versus Female Voice and Voice Changes: Since women generally have shorter and less massive vocal cords than men, the pitch of a woman's speaking voice is higher in pitch than a man's voice (see "Pitch and Size," above). As a child grows, so does the length and mass of the vocal cords, so the pitch of a child's voice changes with increasing age, especially during the growth spurt around puberty when the voice becomes noticeably lower.

Mosquitoes, Bees, and Hummingbirds: Mosquitoes, bees, and hummingbirds can vibrate their wings fast enough to produce audible sounds. This is why we hear them buzzing.

Mice Squeak and Lions Roar: It is of no surprise that mice cannot produce low pitch sounds and lions cannot produce high pitches (see "Pitch and Size," above).

Piano and Harp Strings: Observe the strings on a harp or open up the back of a grand piano and marvel at the variety of string sizes, tensions, and thicknesses (see "Stringed Instruments," above).

Moon Silence and Star Trek: Since there is no atmosphere (no medium) on the moon, sound does not exist. The moon is totally silent. Also, since sound cannot propagate through the vacuum (no medium) of space, don't believe the sounds (explosions) you hear when watching a space battle in *Star Trek* or other space TV shows or movies.

Snowfall and Acoustical Tile: After a new snowfall, everything seems so quiet. Sound does not reflect well off new snow because the many small pockets of air in the snow trap and absorb the sound. The same is true for acoustical tiles (e.g., ceiling tiles) made with fibrous material and holes.

Singing in the Shower: Since sound reflects very well off shower walls and the shower stall is a small and confined space, the reverberation (multiple sound reflections) is significant. This can make your voice sound fuller (and better) than it really is.

Rooms Without Furniture: Listening to sounds in a totally empty room is different from listening to sounds in the same room with furniture and curtains. The multiple reflections (reverberations) in a totally empty room produce a much different sound than the when reflections are reduced due to sound-absorbing furniture and curtains.

Singing Blades of Grass: You can make a blade of grass sing by blowing through a small vertical slit that has been cut into the middle of the blade. See "Squealing Balloons and Boxes," above.

Voices in an Adjoining Room: Sound travels through a wall. You can often hear a TV show or people talking in an adjoining room. The sound travels through both the solid wall material and the air inside in the wall. To reduce such noise pollution, sound-insulating material (absorbing material) can be placed inside the wall.

Interstate Sounds: Often, near interstates and major highways, large walls are erected to block (reflect) the sounds of cars and trucks away from adjacent housing areas.

Thunder and Lightning: Since light travels so much faster than sound, you can use the difference in time between seeing the lightning and hearing the subsequent thunder to estimate how far away the lightning is from your location. Since the light arrives (essentially) instantaneously and the speed of sound is roughly 1100 feet per second, each second of delay translates into about 1100 feet of distance or approximately two-tenths of a mile (0.2 miles). For example, if the delay is five seconds, then the lightning is approximately 5 × (0.2 miles) = 1 mile away.

Sonic Booms: When a jet aircraft flies overhead moving faster than the speed of sound (1100 feet per second or 750 miles per hour), a sonic boom can be heard on the ground. A sonic boom is a large-amplitude sound wave (a shock wave) left in the wake of the aircraft, fully analogous to the V-shaped wake left by a boat moving on water (when the boat is moving faster than the speed of surface water waves). Faster-than-sound travel is now usually restricted to aircraft flying over the ocean to minimize the impact of sonic booms on humans.

Common Sound Intensities: Humans can hear sound intensities (loudness) from as low as 0 decibels (0 dB) at the threshold of hearing to 120 dB at the onset of pain. Above 120 dB, damage to the ear can result. Normal conversation is around 50 dB, a loud truck passing nearby is around 90 dB, nearby thunder is typically 110 dB, and a whisper is about 30 dB. Of course, the intensity will vary with distance from the source. Sound from a jet plane at takeoff can be as high as 130 dB, and this is why baggage handlers and other runway personnel wear ear protection. Factory workers and machinists who work near large and noisy machines are also required to wear ear protection. Loud rock concerts, especially if you are sitting or standing close to the stage and speaker systems, can also cause potential damage to the ear.

Record Player on Different Settings: Old record turntables had adjustable settings for different revolutions per minute or rpm (45, 33-1/3, and 78). If you play a record on a different setting than required, you can change the pitch of recorded music. You can make The Beatles sound like The Chipmunks.

Singing Wine Glasses, Squealing Chalk, Basketball Sneakers, and Tires: In some cases, when one surface slides over another, a squealing sound can be generated. This sound is a result of a slip-and-stick phenomenon in which one surface slides over another for a short distance before it momentarily stops, then slides again, stops, slides again, stops, and so on. If this slipping and sticking happens at a fast enough frequency, it can set one or both of the objects into vibration, which in turn can produce an audible squealing sound. This can happen when a wet finger is slid around the rim of a wine glass, when a piece of chalk is moved across a chalkboard, when basketball sneakers slide across a basketball court, or when you slam on the brakes in a car and come quickly to a stop.

Speed of Sound and Temperature: The speed of sound in air increases with increasing temperature. This is the reason why woodwind musicians always warm up their instruments before tuning them.

Sound Dishes at Sporting Events and Dome Focusing: Sound dishes (bowl-shaped receivers) are used at various sporting events to listen to the players in action. The sound waves coming from the playing field are focused by the dish onto a microphone. The microphone receives and amplifies the sound. In some dome buildings, when you direct your voice upward toward the dome, the reflected sound waves are focused back to you and you hear a loud sound.

WAVES AND SOUND CIRCUS

The following set of activities, selected from the activities described earlier, might be used to initiate your unit on waves and sound. These activities could be set up around the classroom in a circus format—a waves and sound circus. Next to each station, a simple description of the activity is displayed, along with a question or questions to initiate the investigation. Obviously, you will need to rewrite the descriptions and questions given here in order to make the language and analysis appropriate for your grade level. It is suggested that students work in pairs or small groups. Another option would be to use some of these activities (or other activities) as teacher demonstrations to initiate whole-class discussion. Another option would be to have students perform the activities a few at a time and run the circus over a few days. In any case, students should be encouraged to probe the activities beyond the descriptions and initial questions and to think of additional questions they might want to investigate later in the unit.

1. Giant Slinky and Slinky Rope Waves
The giant Slinky and the Slinky rope provide wonderful media in which to make waves and observe a variety of wave properties. Stretch the giant Slinky or Slinky rope (up to 20 feet) along a smooth floor with you at one end and your partner at the other. One at a time, make waves by shaking one end of the Slinky sideways, back and forth along the floor.

Practice making waves of different shapes and sizes. Draw some pictures of some of the wave patterns you make. Be sure to try shaking the Slinky or Slinky rope back and forth just once. What happens to this wave pulse when it gets to the end of the Slinky or Slinky rope? Also draw a picture of what you see when you shake the medium back and forth repeatedly. Repeat, but now shake the medium back and forth more quickly. Increase the tension in the Slinky or Slinky rope by stretching the rope even more to see if this changes how fast your waves move. What do you observe?

2. Water Waves With an Overhead Projector

Turn on the overhead projector and adjust the focus until you can see the water waves in the tray focused on the screen. Produce some water waves by dabbing a pencil tip into the water and watching the screen. What do you observe? Observe the waves reflect off the side of the tray. What is the shape of the reflected waves? Try making two sets of waves by dabbing two pencils into the water a short distance from each other. What happens when the waves cross each other? Do the waves pass through each other? Instead of using pencil points to make the water waves, try dabbing the long edge of a ruler in the water and making "flat" waves. Observe how these waves propagate and reflect. Make two sets of flat waves at an angle to each other and observe them as they cross each other. Do they propagate through each other? Try reflecting the flat waves off the side of the tray. Are the reflected waves flat also? Use the clay to make different shaped barriers in the water (e.g., C-shaped and flat). Send both circular (use the pencil) and flat (use the ruler) waves at these barriers and observe their reflections.

3. Mystery Sounds

Shake each of these containers and listen to the sound. Make a guess as to what is inside the container based on the sound you hear.

4. Tuning Fork Play

Write down what happens in each of these cases:
   A. Pick up a tuning fork and strike it gently against a tabletop. Touch the prongs to the surface of the water in the cup. What do you observe?
   B. Pick up a tuning fork and strike it gently against a tabletop. Touch the prongs gently with your fingers or touch the prongs to a piece of paper. What do you observe?
   C. In a darkened room, set the stroboscope frequency to the frequency indicated on a tuning fork and observe the vibrating prongs in strobe light. Describe what you see.

D. Suspend a Ping-Pong ball from a string. Touch the vibrating fork to the ball. What happens?

E. Bring the vibrating tuning fork close to your ear and listen. Try different tuning forks. How do the sounds differ?

F. Touch the end of the stem, not the prongs, of the vibrating tuning fork to the skull bone behind your ear and listen to the sound that is transmitted through your skull directly to the middle ear. Try different tuning forks. Place the stem of a vibrating tuning fork on the bony part of your elbow, and, at the same time and with the same arm, place a finger in your ear. Can you hear the sound?

G. Place the end of the stem of the vibrating tuning fork (not the prongs) firmly against a tabletop and listen. What do you hear? Try listening to the tuning fork by placing your ear directly on the tabletop.

5. Sound Waves in Liquids

Does sound travel through water? Take the two spoons and submerge them in the water. Click the two spoons together and listen. Can you hear the sound? Now try this. Set the cooking timer to one minute. Place the cooking timer in the plastic bag and seal it tight. Place the bagged timer under the water and wait for the timer to go off. Can you hear the sound of the timer?

6. Sound Waves in Solids

You will need to locate a yardstick and a watch or clock that has a distinctive ticking sound. To investigate sound traveling through a solid, place your ear to the end of the yardstick and hold the ticking watch or clock firmly against the other end. Can you hear the sound? Repeat, but now use a desktop instead of the yardstick. Hear anything? Instead of using a ticking clock or watch as the sound source, try tapping or scraping the yardstick or desktop with your finger. You can also place your ear to a wall and have a friend tap the wall.

7. Microphone and Oscilloscope

Direct the following sounds into the microphone and observe the wave patterns on the oscilloscope:

- Smaller (higher frequency) and larger (lower frequency) tuning forks
- Musical notes of both high and low pitch
- Voiced vowel sounds of higher and lower pitch

What seems to be the main difference in the shape of the pattern between higher and lower pitch for the three cases?

Also, for a given frequency/pitch, what seems to be the difference in the pattern for softer (lower intensity or loudness) and louder (higher intensity or loudness) sounds? In other words, as the sound fades away and becomes softer, how does the pattern change?

8. Pitch and Size

Listen to the sound produced by the following pairs of objects in the following situations:

A. Use the string to suspend two nails of different lengths from the edge of a table. Tie the string just below the head and suspend the nail vertically. Strike each nail with the spoon and listen to the sounds.

B. Pick up one of the metal rods and balance it horizontally by holding it with your fingertips at the middle point. Strike one end of the rod with the hammer and listen to the sound. Repeat with the other rod.

C. Make the smaller and then the larger toy wooden frogs croak and listen to the sound produced by each.

D. Strike the smaller and then the larger tuning fork and listen to the sound in each case.

E. Ring the smaller and then the larger bell and listen to the sound in each case.

What pattern do you notice in these five cases?

9. Garden Hose and Stethoscope

Talk softly into one end of a garden hose while a classmate listens at the other end. Run the hose from one room to another to see if you and your classmate can communicate through the hose. Use the stethoscope to listen to your heart beating. Use the stethoscope to listen to your lungs as you breathe deeply in and out. How are the stethoscope and hose similar?

10. Wok Reflections

Hold the wok at arm's length opposite your face with the inside of the wok facing you. Speak continuously in a normal voice as you very slowly bring the wok toward your face. Describe what you hear. Turn the wok around, with the inside of the wok facing away from you, and repeat.

11. Coat Hanger, Spoons, and Grill Chimes

Wrap a few turns of one end of the string attached to the hanger around the pointing finger of one hand and wrap a few turns of the other end of the string around the pointing finger of your other hand. Insert these fingers in each ear as you dangle the coat hanger below you. Have the dangling coat hanger strike a nearby object (chair, table, etc.) and listen to the sound waves. Repeat using dangling spoons. Try hanging the grill from the string and repeat, but this time have someone stroke across the grill with a spoon. How does the sound get from these objects to your ear?

12. Record Player Fun

Roll the paper into a cone and use the masking tape to hold it together. Stick a sewing needle through the cone about 1/2 inch from the smaller end. The sharp end of the needle should stick out about 1/4 inch. Turn on the record turntable. Now hold the cone such that the needle is resting gently on the rotating record. What do you hear? What do your fingers feel as you hold the cone? Try doing it without the cone and with only a hand-held needle.

13. Cup and String Telephones

To make your telephone, poke a hole in the bottom of two identical cups. Pull a string through the bottom of each cup and tie a paper clip onto the end of the string that is inside the cup so that the string doesn't fall out.

When talking and listening, make sure the string is taut so the sound can be transmitted easily.

Investigate using paper cups, Styrofoam cups, plastic cups, and metal cups (all with kite string) to find out which type of cup makes the best telephone. You might also want to try various strings (all with the same cups) to find out which type of string is the best.

Other fun things you might try:

A. With a single set of telephones and both partners listening, have someone else pluck the string between them. Change the tension in the string and repeat.

B. Hang a metal coat hanger on the string and have someone hit it as two partners both listen through the cups.

C. Loop the lines from two sets of cup telephones around each other. Now each person has a cup for talking and another for listening.

D. Make a party line by looping two or more sets together. Now many people can participate in the conversation.

SAMPLE INVESTIGABLE QUESTIONS

- *Giant Slinky and Slinky Rope Waves:* What happens when wave pulses are generated at both ends of the Slinky and meet up with each other in middle? Do they bounce off each other or do they go through each other?

- *Water Waves With an Overhead Projector:* If the water tray on the overhead is tilted slightly so that one side of the water is at a different depth than the other side, will this affect the motion of the water waves?

- *Tuning Fork Play:* Can you amplify the sound of a vibrating tuning fork by touching the stem to other objects in the classroom? Which objects amplify the sound best? What happens when the fork is touched to the body of a musical instrument like a guitar?

- *Microphone and Oscilloscope:* When you direct two different sounds into the microphone simultaneously, what does the resulting wave pattern on the oscilloscope looks like?

- *Pitch and Size:* What objects in the classroom when struck (gently) by a drumstick give off higher pitch sounds and which objects give off lower pitch sounds?

- *Coat Hanger, Spoons, and Grill Chimes:* What other objects can be hung from the string to produce interesting sounds?

- *Record Player Fun:* How do the size, shape, and material of the cone affect the sound? Is a cone even needed? Can you hear the sound with a handheld needle? What other pointed objects would also work as the needle (fingernail, safety pin, thumbtack, etc.)? What happens to the sound when you have the turntable turning on the other settings (45 or 78 rpm)?

Light and Electromagnetic Waves

WAVES AND/OR PARTICLES

Philosophers and scientists have been debating the age-old question, "What is light?" since Greek philosophers in the fourth century B.C. first speculated that light might be made of "fast-moving particles" that travel in straight lines. Leonardo da Vinci (1452–1519) was the first to speculate that light might have a "wave" character. He came to this conclusion based on a simple yet powerful analogy between sound and light. He proposed that since sound is a wave and sound reflects (echoes; you can hear your voice return after reflecting from a canyon wall), then light might be wavelike, since it also echoes (reflects; you can see your image reflected in a mirror or smooth lake).

The particle-versus-wave debate on the nature of light heated up in the 17th century. As a particle proponent, Isaac Newton (1643–1727) could explain many of the properties of light (image formation, shadow formation, reflection, refraction, dispersion, colors, and polarization) through a particle description. But two contemporaries of Newton, Robert Hooke (1635–1703) and Christian Huygens (1629–1695), argued that these same properties of light were better explained using a wave description. Based more on Newton's reputation and less on the evidence, the particle model of light remained in favor until the early part of the 19th century.

In 1803, Thomas Young (1773–1829) attempted to convince the scientific community that Hooke and Huygens were correct. He performed some stunning experiments (see "Diffraction and Interference," below) that could be explained only through a wave model for light. Through these experiments, he was the first to calculate the wavelengths of different colors of visible light. Still, the scientific community was not willing to give up on Newton's particle model. Indeed, it would take scientists another 15 years before the wave model was fully accepted.

The turning point happened in 1818 when Augustin Fresnel (1788–1827) proposed a mathematical wave model for light. Simeon-Denis Poisson (1781–1840) used Fresnel's mathematical model to predict "an absurd bright spot" in the center of the shadow region cast by a very small object. Poisson, a particle proponent, did the calculation only to show the absurdity of Fresnel's model, but soon the spot was detected experimentally and ironically became known as Poisson's spot. At least for a time, this historical episode put the nail in the coffin of the particle model.

In 1860, in a brilliant mathematical synthesis of the equations of electricity and magnetism, James Clerk Maxwell (1831–1879) theoretically predicted the existence of an *electromagnetic wave* that traveled at a wave speed close to the experimentally known speed of light. The conclusion was obvious. Light is a wave of propagating electric and magnetic fields.

Maxwell's theory also predicted the existence of nonvisible light. The mathematics of the theory did not exclude any wave frequencies, so he concluded that other nonvisible forms of electromagnetic radiation were possible. Twenty-six years later the actual existence of nonvisible light waves would be verified by Heinrich Hertz (1857–1894). In 1886, Hertz experimentally produced and detected nonvisible electromagnetic waves.

So, by the end of the 1800s, all seemed secure. Light, both visible and nonvisible, was a wave, and electric and magnetic fields were doing the waving. But this was not going to be the end of the story, not by a long shot.

In the early 1900s, the electromagnetic wave picture of light failed to explain a number of specific phenomena that were being explored at the atomic level. The problems arose not with the propagation of light in space, but only when light was being produced by matter or when interacting directly with matter. Albert Einstein was one of the first to analyze one of these phenomena, called the photoelectric effect, an effect that involves light striking and ejecting electrons from a metal surface. A pure electromagnetic wave picture of light could not explain the details of the effect—the number and speed of electrons ejected. Only when Einstein proposed that the electromagnetic waves interacted with the metal in a "particle-like" way was he able to explain the effect. While he continued to picture light as a propagating electromagnetic wave, he speculated that the energy and momentum carried by the wave must come in discrete chunks. These discrete chunks or packages of energy and momentum, now called *photons*, interacted with the electrons in the metal in a one-on-one fashion. The number of photons carried by the wave, he proposed, was proportional to the amplitude of the wave, and the energy and momentum of each photon was related directly to the frequency of the wave. So the particle picture of light was back on the scene. Indeed, the nature of light turned out to be much more interesting than an either-or particle or wave reality. Both descriptions are needed for a full picture of light. This complementary picture is known as the *wave-particle duality* for light. The dual nature of light, propagating as an electromagnetic wave while interacting with matter through particle-like photons, sums up our present understanding of light.

This wave-particle duality is now known to be a universal property of nature. Based on an analogy with the dual nature of light, Louis de Broglie (1892–1987) proposed that matter itself (electrons, atoms, etc.) might show the same duality. He proposed that *particles* such as electrons might have wave-like features. His ideas were soon put to the test and vindicated when experiments revealed the diffraction and superposition of electrons!

All forms of light, both visible and nonvisible, share many fascinating characteristics and properties. Light does not need a medium through which to propagate. It can propagate in a vacuum. Light is made up of oscillating and self-sustaining electric and magnetic fields that interact with matter in a particle-like way (photons). Light travels at a constant speed in a vacuum (299,792,458 meters per second). In a transparent material, light travels at a slower speed, with different frequencies traveling at slightly different speeds. In a vacuum and in uniform transparent materials, light travels in straight lines. Some objects reflect light, some absorb light, and some do both. Light bends—*refracts*—when it propagates from one transparent medium into another. Visible light comes in a narrow but continuous spectrum of frequencies that we detect as different colors (red to violet). "White" light is composed of light of different colors (red to violet). "Black" is the absence of light. Light waves *diffract* (bend around objects and spread out through small openings) and obey the superposition principle (when two or more light waves are propagating in the same space, the net effect is simply the sum of the waves). Light waves (photons) pass through each other.

SOURCES OF LIGHT

Concepts

The source of all electromagnetic waves (visible and nonvisible light) is either *accelerating* electric charge (i.e., electric charge that is speeding up, slowing down, and/or turning) or electric charge undergoing quantum transitions within atoms, molecules, and nuclei. Since electric charges (mostly electrons in atoms) are found in all the states of matter—solids, liquids, and gases—*all matter emits light.* The frequencies and intensity of light emitted by matter depends on the particular type of matter (atoms) and the temperature. This makes sense because temperature is a measure of the average speeds of the atoms that compose the matter. At higher temperatures, the faster moving atoms strike each other harder and more often, which produces more "excited" atoms, atoms in higher energy states. As the atoms de-excite to lower states, they emit light (photons). At typical room temperatures, all objects emit nonvisible light (mostly infrared). Hotter objects (stars, fire, lightbulb filaments, etc.) emit visible light (as well as nonvisible light). The hotter the object, the more light it emits and the higher the frequencies of light emitted. Some objects can emit visible light even at lower temperatures through unique electrical and/or chemical processes: the aurora borealis, lasers, computer or TV monitors, neon and fluorescent lights, fireflies, glowing fish, and light-emitting diodes (LEDs), to name only a few.

Nonvisible light, such as radio waves and microwaves, can be produced by a changing current (accelerating and decelerating electrons).

The alternating current in your home wiring produces low-intensity electromagnetic waves at a frequency of 60 Hz, matching the source frequency (60 Hz alternating current). The oscillating current carried in high-power transmission lines emits electromagnetic waves that some claim propose a health threat for nearby residents and animals. Remote-control units emit infrared light that can be detected by electrical receiving devices in your car, home, and garage. A variety of devices can produce electromagnetic waves and detect their reflections (radar, remote sensing devices for mapping the earth and other planetary surfaces, metal detectors for locating lost items, airport and store security detectors, and motion sensors that turn on and off water faucets and flush toilets in public bathrooms, to name only a few).

At the subatomic scale of fundamental particles (quarks and anti-quarks and leptons and anti-leptons), photons of light can be produced through electromagnetic interactions between charged particles or between charged particles comprised of these fundamental particles. For example, a beam of fast-moving electrons (leptons) can produce X-ray photons during the process of undergoing a rapid deceleration into a metal plate (this is how medical X-rays are produced). A lepton (say, an electron) and an anti-lepton (say, an anti-electron, called a positron) can interact with each other, annihilate each other, and produce two photons in the process. There are many of these photon-producing processes happening in particle accelerators around the world, through cosmic ray interactions with gases in the upper atmosphere of the earth, in the interior of stars, in colliding stars and galaxies, and in the early universe just after the big bang. In fact, we are now detecting very long wavelength (low frequency) photons, remnants from the early universe. The photons are of very long wavelength because space has been expanding since the big bang.

Activities

Visible Light Walk: Take a stroll inside and outside the school and have students discover as many sources of visible light as possible. These might include incandescent lights, fluorescent lights, an overhead projector, a laser pointer, lights flashing on a phone (LEDs), flashlights, TV and computer monitors, sun and stars, streetlights, neon signs, car lights, match and candle flames, house lights, light from a cigarette, and laser beam price detectors in grocery stores, among others.

Virtual Nonvisible Light Walk: Take an imaginary stroll inside and outside the school to discover as many sources of nonvisible light as possible. These would include *all* matter—chair, house, human body, and so on.

Specific examples might include radio and TV station towers, cell phones and towers, metal detectors, remote-control units for garage door openers and home entertainment equipment, remote key entry-lock devices, wireless computer hubs, microwave sources (magnetrons and klystrons), X-ray tubes, radar sources at airports and inside airplanes, radar guns for detecting speeding cars and moving baseballs, black lights, satellite sources for GPS and remote sensing work, and cosmic sources of nonvisible radiation.

Visible and Nonvisible Light Sources: Have students bring in examples of sources of light and create a display of them in the classroom. Divide the display into visible and nonvisible sources. Visible examples might include various types of lightbulbs, LEDs, flashlights, laser pointers, and candles. Nonvisible examples would include TV remotes, garage door openers, key-lock remote devices, motion sensors, cell phones, microwave ovens, and metal detectors.

Everyday Examples

Visible Electromagnetic Waves:

Visible light: red, orange, yellow, green, blue, indigo, and violet

Starlight (including our sun)

Fire

Lightning

Incandescent lights

Fluorescent lights

Neon, sodium, and other gas lights

Lasers

Light-emitting diodes (LEDs)

Fireflies

Fireworks

Explosions

Gunfire

Sparks

Flashbulbs

TV and computer monitors

Cigarettes, cigars, and pipes

Matches

Cigarette lighters

Candle flames

Glowing stoves

Glowing burners

Glowing wires in toasters

Glowing wires in electric heaters

Glowing lava

Aurora borealis

Shooting stars

Nonvisible Electromagnetic Waves:

All objects emit nonvisible light

Ultraviolet

Infrared

Microwaves

Gamma rays

X-rays

Radio waves

TV

Radar

Remote controllers (car entry-locks, garage door openers, TV channel and volume changers, etc.)

Animal tracking devices

Tracking, controlling, and communicating with space probes and space vehicles

Cell phones and towers

Microwave transmission

Radiotherapy

MRI (magnetic resonance imaging)

Power-line radiation

Wireless hubs and wireless computers

Wireless phones and wireless microphones

Nonvisible light from stars and other galactic sources (X-rays, microwaves, radio waves, etc.)

CB radios

Satellite signals for GPS and remote sensing

Ground-penetrating radar

LIGHT AND MATTER

Concepts

Matter is constantly emitting light, mostly nonvisible light. Likewise, matter is constantly absorbing light, again, mostly nonvisible light. In fact, when an object is in *thermodynamic equilibrium* with its environment, meaning that the object is at the *same* temperature as its surroundings (like most nonliving objects in a room), the rates of emission and absorption are identical.

Objects emit and absorb different frequencies of light, depending on the types of atoms and molecules that constitute the object. In other words, a given material will emit and absorb only unique frequencies associated with the types of atoms and molecules composing the material. Indeed, this is the way scientists identify most of the material in the universe, by measuring the frequencies of light sent to us from various objects and comparing those light frequencies to that of the light emitted by known substances.

Those materials that have atoms and molecules that absorb and quickly re-emit visible light are said to be *transparent*. The atoms and molecules in transparent objects (glass, for example) quickly re-emit the light before they have a chance to interact with neighboring atoms and molecules, so do not dissipate the absorbed light as heat (atomic/molecular motion). In this way, the visible light is passed from atom to atom, unaffected in frequency but delayed in time. In fact, these atom-to-atom delays lead to a slower average speed of light as it traverses the medium. It is important to note that light still travels at its set vacuum speed *between* atoms.

An object transparent to visible light might not be transparent to other, nonvisible frequencies of light. For example, glass is transparent to visible but not to infrared and ultraviolet light. *Opaque* materials absorb visible light with little re-emission. The energy is transferred to atomic and molecular motion (heat) before it can be re-emitted. Some opaque objects, like metals, are good *reflectors* of light. In metals, some of the electrons are more or less free to move throughout the metal (this is why metals are good conductors of both electricity and heat). These "free" electrons are set into vibration by the light incident upon them. Since they are unattached to the metal atoms, these vibrating electrons cannot transfer these vibrations to the surrounding atoms. The light is simply re-emitted as reflected light.

Activities

Transparent Objects 1: Locate objects in the school that are transparent, objects through which students can actually see each other's images. These might include glass and clear plastic windows, eyeglasses, water in a fish tank, various liquids in a glass, clear plastic rulers, glass and clear plastic boxes and coverings, glass and clear plastic bottles and jars and bags, clear plastic wrap, lenses, clear plastic tape and laminates, and the like.

Transparent Objects 2: You will need to locate a flashlight and a laser (Standard Model Laser Pointer, Arbor Scientific). Locate objects in your classroom that are not transparent enough to pass an image but through which some light can pass. Test these objects in a darkened room by seeing if some of the light from a flashlight or the laser held on one side of the object will pass through to the other side. Try various paper and plastic products, clothes, fingertips, tape, frosted glass, and so on.

Thick Versus Thin Transparency: You will need to locate a ream of paper (or a book), a flashlight, and a laser (Standard Model Laser Pointer, Arbor Scientific). Most thick materials that are opaque to light become partially transparent when cut thin. In a darkened room, using either the flashlight or a laser, determine how many sheets of paper you need to block out all light. You can also use the paper pages of a book to do this activity. Try different types of paper or other materials. For example, a laser beam will not be passed through the thickest part of your hand but can be partially transmitted through thinner parts.

Everyday Examples

Transparent Objects: Objects transparent to visible light are all around us. A short list would include glass and clear plastics, water and other clear liquids, and air and most other gases.

Opaque Objects: Objects opaque to visible light are all around us. In fact, most things in your classroom, including you, are opaque to visible light.

LIGHT TRAVELS IN STRAIGHT LINES

Concepts

In a vacuum, light travels in a straight line, ignoring the very small effect that gravity has in curving a beam of light. In most cases, light travels in a straight line in transparent media. This is true if the medium is of uniform density (same density throughout the medium). This is generally the case for common media such as air, water, and glass. But if the transparent medium has a density gradient, the light will follow a curved path. For example, light coming from the sun and traversing the earth's atmosphere actually follows a curved path, a direct result of the fact that the air density is greatest near the surface of the earth and thins with higher altitude.

Activities

Important Notes: To perform most of the activities described in this and subsequent sections, a darkened room is essential, the darker the better. Unless your classroom comes with light-darkening shades, you will need to cover all windows with dark paper or some other opaque material.

You will also need to locate or construct light sources that concentrate light into a beam. Four are suggested here:

Laser: Cheap laser pointers are now available commercially (Standard Model Laser Pointer, Arbor Scientific), and you can purchase a tripod stand to hold the laser (Laser Tripod, Arbor Scientific). Lasers provide a very concentrated beam of light and are ideal for many of the following activities. Of course, caution must be followed in using any laser, so read the warning label closely. Students should use lasers only under controlled situations with direct adult supervision.

Modified Flashlight: Flashlights produce light in a widening beam, useful for seeing things in the dark, but less useful as a concentrated beam of light. A flashlight can be adapted to form a more concentrated beam by taping over most of the opening with black electrical tape or duct tape, leaving only a small, square, quarter-inch opening in the center.

Modified Slide Projector: Cut out a square of thin cardboard (e.g., from the backing of a note pad) to the size of a projector slide. Cut a small, square, quarter-inch opening in the middle of the cardboard slide. Place this slide in the projector. This will help concentrate the projector light into a narrower beam.

Modified Overhead Projector: Construct a cardboard mask and place it on top of the overhead projector. The mask needs to cover the entire overhead projector's light surface, except for a small, square, quarter-inch opening in the middle of the mask. This will help concentrate the overhead light into a narrower beam.

For many of the following activities you will need to make the light visible in the air. Both Educational Innovations (Diffusion Mist) and Arbor Scientific sell a can of spray that emits a nontoxic aerosol fog for this purpose (no more chalk dust!). You can also put chalk dust in the air, but the spray fog is safer.

See the Light Beam: You will need to locate the four light sources and spray fog as described above. In a darkened room, shine the laser light across the room and spray the fog into the beam. Notice that the light is traveling in a straight line. Spray the fog farther and farther down the beam of light to expose the entire beam.

Let There Be Light: You will need to do this activity on a sunny day when sunlight is beaming into the classroom. You will also need the fog spray. In a darkened room, allow light through only a small part (e.g., one square foot) of the window through which the light is entering. The beam should cast a rectangular-shaped bright spot on the floor. Notice that the dust in the room will make the beam somewhat visible. To see the beam more clearly, spray the fog in the vicinity of the beam. Students might also be interested in recording the movement and shape changes of the bright spot on the floor over the course of the day. If sunlight does not enter your classroom directly, you might try using a mirror that you have positioned outside your window to reflect the sunlight into the classroom.

Beam Box: You will need to locate a cardboard box, a small lamp, a bright and clear lightbulb, and the fog spray. For best results, the lightbulb should be of high intensity and unfrosted; that is, the bulb is made of clear glass and you can see the filament. Place this bulb in the lamp, turn on the lamp, and cover the lamp with the cardboard box. Punch a few small holes in the box to let some beams of the light escape. In a darkened room, notice where the light hits the walls. To "see" the beams emerging from the box, spray the fog around the box. Notice that the light radiates out from the lightbulb in straight lines.

Line of Cards: You will need to locate some index cards and some clay. You will also need the laser and modified flashlight sources. Cut a dime-sized hole right in the middle of 10 index cards. On the surface of a long smooth table, stick 10 wads of clay a few inches apart along the length of the table. These wads serve as holders for the index cards. Insert an index card in each wad, lining them up like dominoes across the table. The idea now is to shine the laser beam so that it successfully goes through the hole in all 10 cards. You will need to adjust the cards in order to obtain this result. Once you have succeeded, see if you can get the light from the flashlight source to successfully traverse the holes. Now, ask students if they can see each other through the "line" of holes. Light travels in straight lines.

Indoor Shadow Play: You will need to locate an overhead projector (unmodified) or slide projector (unmodified) and large sheets of white paper. Attach the paper to the blackboard or classroom wall. In a darkened classroom, using either the overhead or slide projector light source, have students cast shadows of objects and/or of themselves onto the paper. Students can draw outlines of the shadows on the paper. Try drawing classmates' silhouettes. Try designing shadow puppets and putting on a shadow play. Cast shadows of "unknown" objects (hidden behind a screen) to see if students can guess the object. Students should also be encouraged to investigate how the shadow's size changes with the distance the object is from the paper and/or from the light source.

Shadows on Earth: You will need to locate a globe and a light source (unmodified overhead or slide projector). Attach a very small toy figure (or some other very small object) to the globe at the approximate location of your state. In a darkened room, hold the globe in the projector light and investigate the bright and dark side of the globe (day and night). Rotate the globe to simulate day going into night and night into day. Observe the shadow cast by the figure as the globe is rotated in the light. Notice that the longest shadows cast by the figure on the globe's surface occur when the figure is approaching the dark to light boundary near the edge of the globe (at sunset or sunrise) and that the shortest shadows occur when the figure is in the middle of the light (at noontime). This simulation also provides a good way to illustrate how the sun rises in the east and sets in the west and the reason for time zones. By tilting the globe on its axis, you can also simulate the reasons for seasons: longer summer days, shorter winter days, and the lack of a setting sun in extreme northern and southern latitudes. All of these effects are a direct result of the fact that light travels in straight lines.

Outdoor Shadow Investigation: You will need a bright day in the fall, one in the winter, and one in the spring. You will also need some yardsticks and large pieces of paper. Locate a flat portion of the playground. With students working in pairs, one student stands on a piece of paper and casts his or her shadow over the rest of the paper while the other student outlines the full silhouette on the paper. The student should also be supporting a vertical yardstick that is touching the ground near his or her feet. This is repeated for the other student in the pair. These shadow silhouettes are made at three times during the day—at a set time in the morning, at noon, and in the afternoon. Students should measure and record the length of their shadow and the length of the shadow of the yardstick for each of the times during the day. This procedure should be done three times during the year—fall, winter, and spring. Analyze the results.

Handheld Pinhole Camera: You will need a bright light source (lightbulb), a candle, a large paper cup, aluminum foil, wax paper, and rubber bands. Remove the bottom of the paper cup and replace it with aluminum foil. You can fix the aluminum foil to the bottom with either a rubber band or tape. Cover the mouth of the paper cup with wax paper, fixing it with either a rubber band or tape. Using a pen or pencil, punch a small hole in the middle of the aluminum foil. Light the candle and darken the room (the darker the better). To test your pinhole camera, hold it a few feet away from the candle flame, pointing the pinhole side toward the candle flame. Look at the wax paper. You should see an image of the candle flame upside down and reversed left to right. Repeat, but this time use the bright light source as your object. The pinhole camera is based on the fact that light travels in a straight line, since that is the only way that light emitted from the top of the candle flame that makes it through the pinhole can strike the lower part of the wax paper. For the same reason, light from the bottom of the flame that goes through the pinhole must strike the top part of the wax paper. Continuing this reasoning, light from the left part of the flame that goes through the pinhole must strike the right side of the wax paper and the light from the right part of the flame must strike the right side of the wax paper. So, indeed, the image should be upside down and reversed.

How does the image change when the pinhole camera is moved closer to or farther from the light source? What happens when you punch more than one hole in the aluminum foil? Which works better, a pinhole camera with a large or small pinhole? How does the length of the pinhole camera—the cup—affect the image?

Instead of using a paper cup, you might try to make a pinhole camera out of a cereal box, a cone made from black paper, or a soup can.

The Classroom as a Pinhole Camera: You will need a classroom that is very dark and a very bright day outside. Make sure that all the light leaks in the classroom are sealed. In the middle of one of the sealed windows, you will need to establish a small opening (the pinhole, an inch in diameter). On the wall opposite the pinhole window you should observe a beautiful image of the entire scene outside the classroom. It will be upside down and reversed. Have someone go outside and walk left to right across the scene and watch that person walk upside down and right to left across the classroom wall.

A Big Box as a Pinhole Camera: Another version of the pinhole camera activity uses a large box (like the big box a refrigerator is shipped in) instead of the classroom. A student stands inside the dark box. The box has a pinhole in one wall and a white sheet of paper on the inside opposite wall. With the box outside on a bright day or with the pinhole facing a window, the student inside the box will see an upside-down and reversed image of the scene on the white paper.

Fuzzy Shadows: You will need to locate a smooth ball about the size of a tennis ball, a light source (a lamp with lightbulb), and a screen. The lamp shade should be removed from the lamp. Hold the ball relatively close to a screen. Darken the room except for the light source. Observe the shadow of the object on the screen. Bring the light closer to the ball and notice what happens to the shadow. Because light travels in straight lines, a fuzzy shadow, the *penumbra*, forms around the darker shadow in the center, the *umbra*. In the penumbra, the light from one side of the light source is blocked from reaching the screen, while the light from the opposite side of the light source is not; thus, the shadow here is not as dark. In the umbra, all the light is blocked from reaching the screen.

Shadows: You will need to locate a small light source (shorter filament lightbulb), an extended light source (longer filament lightbulb), and a screen. In a darkened room, turn on the small light source. With your hand near the screen, cast a shadow of your hand onto the screen. Move your hand toward and away from the light source. Repeat with the extended light source. What is different about this shadow compared to the shadow formed from the small light source? See "Fuzzy Shadows" activity, above.

Everyday Examples

Blind Corners: We cannot see around corners because light travels in straight lines.

Hidden Objects: We cannot see objects hidden behind other objects because light travels in straight lines.

Locating Objects: We have the ability locate objects because light travels in straight lines to our eyes.

Laser Beams: Laser beams are used in mapping out straight lines on a construction site, in locating straight lines in surveying, in surgery, to guide missiles, as a screen pointer during presentations, and in laser shows, to name only a few applications.

Shadows: All shadows result because light travels in straight lines.

Hand Puppets: Children enjoy making shadow hand puppets when a bright light is shining on a wall or screen.

Height Measurement: Since light travels in straight lines, by measuring both the *length* of the shadow of an object (a tree, building, etc.) on the ground and the *angle* that the top of the object makes with the ground (sighted from the end of the shadow), you can determine the height of the object. The height can be calculated by multiplying the shadow length times the tangent of the angle.

Distance Measurement: Since light travels in straight lines, by measuring the difference in angle that results when an object is viewed from two spatially separated points (a known distance apart), you can determine the distance to the object. This powerful technique, known as *parallax,* is used to determine distances in surveying, distances to aircraft, as well as the distances to astrophysical objects (planets, asteroids, and stars).

Seeing Depth: Having two spatially separated eyes allows us to perceive depth in our three-dimensional world. See "Distance Measurement," above.

Fuzzy Shadows: An object illuminated by an extended light source (lightbulb, fluorescent light, etc.) will cast a fuzzy shadow (see the "Fuzzy Shadows" activity, above).

Night and Day: Night is dark because the sunlight, which is traveling in straight lines, is blocked by the earth.

Eclipses: Eclipses are formed by light being blocked by the earth (lunar eclipse) or by the moon (solar eclipse).

Phases of the Moon: The phases of the moon are caused by light striking the moon at different times in its monthly orbit around the earth: new, new crescent, first quarter, waxing gibbous, full, waning gibbous, last quarter, and old crescent.

Observing the Sun: Since you are never to look directly into the sun, a good way to view the sun is by punching a small hole in a 3"-by-5" card and holding the card close to the ground above a piece of white paper. You will see a pinhole image of the sun on the paper. This is a good way to view a solar eclipse safely.

Pinhole Leaves: Sometimes you can see multiple images of the sun on the ground below a tree. The small openings between the leaves can act as pinholes and produce the multiple images. During a solar eclipse, they change in shape from round to crescent and back again.

Artists and Pinhole Images: Before the invention of the photograph, some artists used pinhole cameras (camera obscura) to image scenes they were painting.

Buddha Rays: Sometimes sunlight traveling between clouds produces a beautiful "fan" of light from the clouds to the ground. The light rays from the sun arrive essentially parallel to each other. The fanning effect is just a matter of perspective, analogous to a road or railroad track appearing to narrow in the distance.

Smoke, Fog, Snow, and Rain: Car headlights or flashlights can produce visible beams of light when they shine through smoke, fog, snow, or rain.

Light in a Dusty Room: When sunlight from a window shines into a dusty room, you can sometimes see the beam in the room. The same is true in dusty theaters with movie projectors or in a dusty classroom with overhead or slide projectors.

Searchlights: Sometimes you can see the light beams from giant searchlights that are used to advertise the location of a big event.

Evolution of the Eye: The eye has evolved to see and locate objects. To locate objects, the eye-brain "assumes" that the light entering the eye has been traveling in a straight line. But the eye-brain can be fooled. For example, when light from an object is reflected or refracted before it enters the eye, the eye-brain still assumes a straight-line path and, as a result, some interesting optical illusions can be created (i.e., objects are not where they appear to be).

Radio and Cell Phone Towers: Since radio waves travel in straight lines, large structures, mountains, and the curvature of the earth can limit their range. To extend the range, radio and cell phone towers are often placed on the top of mountains or tall buildings.

Remote Controllers: You must aim the remote controller at the TV set for it to work.

Alignment of Satellite Dishes: Since light travels in straight lines, satellite dishes must be aligned accurately, aimed directly at an orbiting satellite, in order to receive the maximum intensity of the electromagnetic waves.

Satellite Radio Shadows: Buildings, bridges, and mountains can cast shadows on radio waves from orbiting satellites. The reception of satellite radio in your car is momentarily blocked when you are in the "shadow" of such structures.

REFLECTION

Concepts

You can see your image in a mirror, a smooth lake, a shiny car, or a store window, but you cannot see your image at all in a cement wall or piece of paper. Light reflects very differently from different materials. Light incident on a smooth surface, like a silvered mirror, is not dispersed in all directions after it strikes the surface, but is *reflected at an angle equal to the incident angle.* In other words, as shown in Figure 4.1a, the incident angle A is equal to the angle of reflection B. This is known as the *Law of*

Figure 4.1a

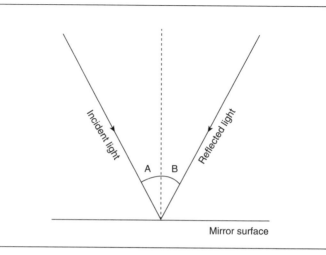

Mirror surface

Figure 4.1b

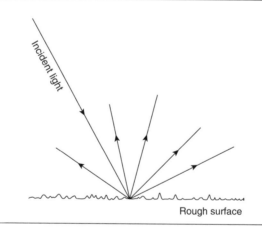

Rough surface

Reflection. On the other hand, light incident on a rough surface, like a piece of paper, is dispersed in many directions upon reflection, as shown in Figure 4.1b.

The mirror-type reflection is called *specular* or *mirror* reflection and the paper-type reflection is called *diffuse* reflection. Diffuse reflection is by far the most common type of reflection. Indeed, most objects in the world around us (you, me, furniture, walls, trees, moon, planets, etc.) disperse the light they reflect. This type of reflection allows us to see such objects from many directions. On the other hand, some smooth materials, especially polished metals, do not disperse the light incident on them, but simply change the direction of the incident light according to the Law of Reflection. As we will see, the Law of Reflection allows us to understand many intriguing phenomena, including image formation and light focusing.

In order to see how image formation follows directly from the Law of Reflection, consider a point P on the surface of an object (say, an arrow) from which light is emanating (due to diffuse reflection) in straight lines in all directions. Now place a *flat mirror* nearby. Some of the light from the point P, spreading out in straight lines, will strike and reflect off the mirror. According to the Law of Reflection, each ray will reflect off the mirror at the same angle it hits. This construction is shown in Figure 4.2a.

In Figure 4.2a, note that even though the source of the light is really point P on the object, the reflected light forms its own bundle of rays that would be identical to a bundle of light coming from an image point I behind the mirror. In Figure 4.2b, an observer's eye has been included. The eye receives only part of this bundle of rays. In fact, the circular eye opening receives a bundle of rays in the form of a cone of light that appears to be coming from the image point I. Since your eye-brain assumes that the light has always been traveling in straight lines, you, the observer, would conclude that this cone of light was emanating from the image point I at a depth *inside* the mirror exactly equal to the distance the object point P is in front of the mirror. Remember, your eye-brain thinks that the light has always been traveling in straight lines, so it is fooled and "sees" the object point P inside the mirror at the image point I. The same is true of light coming from all other points on the object and reflecting off the mirror and entering the eye of the observer.

To extend this analysis, now consider two points on the object, P(1) and P(2), as shown Figure 4.3, and follow the cone-shaped bundles of light coming from each of these points that strike the flat mirror and enter the observer's eye.

The two image points, I(1) and I(2), have been determined, again, by obeying the Law of Reflection and assuming the eye is fooled by the always-in-a-straight-line trick. Note that the size of the image, represented by the distance between points I(1) and I(2), is the *same size* as the object, represented by the distance between points P(1) and P(2). But since the image is a given distance from the observer (actually at a distance double the distance from the observer to the mirror), normal perspective will make the image look smaller, just like any object looks smaller the farther away it is. Also note that the image is *right side up* (not flipped upside down) and *always seems to be as deep within the mirror as the object is in front of the mirror.* Interestingly, the image will appear *rotated* compared to the object. For example, as you face a flat mirror, your image is looking back at you as if your body has been rotated around it vertical axis. To see

Figure 4.2a

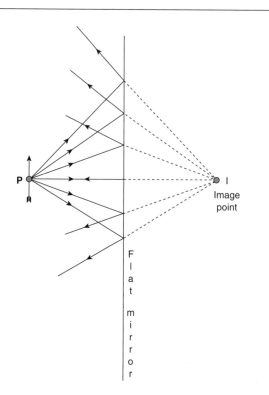

Figure 4.2b

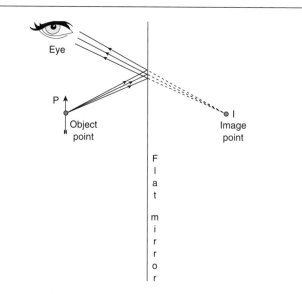

Figure 4.3

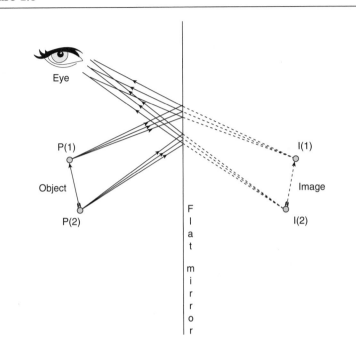

why this happens, imagine that the point P(2) in Figure 4.3 is a point on your right hand as you face and look into the mirror. Since the image point I(2) will be located to the image's left inside the mirror, you will see your right hand imaged as a left hand inside the mirror.

Images produced by curved mirrors are even more fascinating. To begin the analysis, consider two points on the object, P(1) and P(2), as shown in Figure 4.4. Follow the bundles of light coming from each of these points that strike and reflect off a *convex mirror* (mirror surface is bowl-shaped and curved outward) and enter the observer's eye.

As always, the Law of Reflection has been applied to all of the rays used in the construction. Note the difference in the reflected bundles compared to the flat mirror. Each of the reflected bundles of rays spreads out more quickly because of the curved surface, and, as a result, the image points, I(1) and I(2), found by assuming the rays have always been traveling in straight lines, are *not as deep* within the mirror as compared to the image in a flat mirror. Also note that the size of the image, represented by the distance between points I(1) and I(2), is *smaller* than the size of the object, represented by the distance between points P(1) and P(2). The image is *right side up* and *located more toward the center of the mirror.* The image will also be *rotated,* similar to the case for the flat mirror.

Figure 4.4

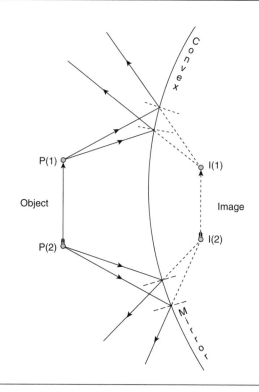

It is an interesting mental exercise to imagine how an image would change if a flat mirror were slowly warped into a more and more convex one. As you continue to warp the mirror, the image will continue to shrink in size, come closer and closer to the mirror's surface, and move more toward the middle of the mirror. The mirror will also be able to sample a larger and larger physical space as it gets more and more convex, since the curved parts of the mirror can now reflect light from objects that were not possible to see with a flat mirror. Indeed, this means that the image in a convex mirror concentrates a larger object space into a smaller image space.

Next, we will consider the even more fascinating images formed by a *concave mirror*, where the mirror surface is bowl-shaped and curves inward. As we will see, two very different types of images can be formed in such a mirror, depending on whether the object is placed close to or far away from the mirror surface. To begin, consider an object close to the mirror surface, as shown in Figure 4.5, and note how the light bundles, emanating from two points P(1) and P(2) on the object, reflect off the mirror surface.

As always, we constructed the result based on the Law of Reflection. Notice the locations of the image points, I(1) and I(2). In this case, the

Figure 4.5

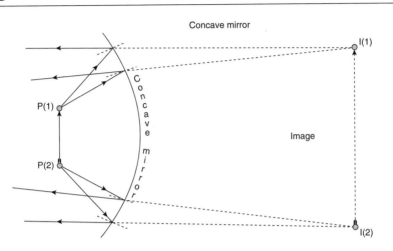

Concave mirror

image will be *right side up* and *larger* than the object. The image will also be *deeper* inside the mirror than the object is in front of the mirror.

But the results change when the object is farther away from the mirror. In Figure 4.6, we can see how this image change is produced.

Now each of the ray bundles from points P(1) and P(2) on the object, after reflecting off the mirror, actually *crosses* the others at a point in front

Figure 4.6

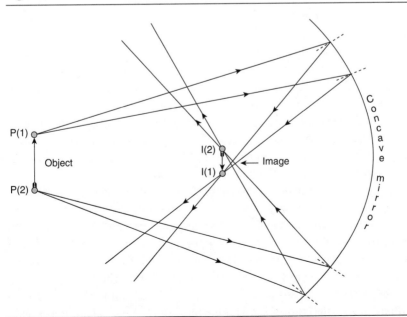

Figure 4.7

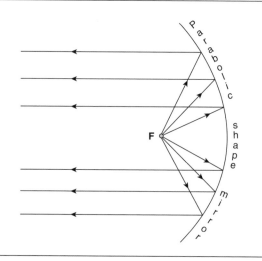

of the mirror. Consequently, an observer's eye, seeing this bundle of rays, believes that they have originated from these crossing points, image points I(1) and I(2), *in front* of the mirror. Also note that the image is *upside down* and *smaller* than the object. This kind of image is very different from the images formed by flat and convex mirrors or for objects placed near a concave mirror. In these earlier cases, real light did not actually pass through the image points; indeed, those image points were behind the mirror surface. In this new case, however, light actually passes through the image points in front of the mirror. Such an image is called a *real image*. A *virtual image* is one for which light does not actually pass through the image points, as in the flat, convex, and near concave cases.

So what happens to the image of an object in a concave mirror when it or part of it is at the particular point in front of the mirror where the transition occurs between these two near and far cases? As shown in Figure 4.7, there is a unique point F, called the *focal point*, where the rays reflecting off the mirror never cross (for a parabolic-shaped concave mirror), neither in back of the mirror as in the case of near concave nor in front of the mirror as in the case of far concave.

From this unique point, the rays reflect off the mirror *parallel* to each other. The image gets spread out and becomes unrecognizable. It is interesting to watch the image of an object in a concave mirror change as you move the object from close to the mirror and then out through the focal point. Close to the mirror, the image is large and right side up. As the object is moved away from the mirror and toward the focal point, the image continues to grow in size. At the focal point, the image has expanded so much that it is just a big blur. Once the focal point has been traversed, a smaller, upside-down image suddenly appears, one that gets smaller and smaller the farther the object is from the mirror.

A concave mirror is often used to focus light. To see how this is done, simply reverse the arrows in the Figure 4.7. A concave mirror can also be used to create a beam of light. This can be done by placing a light source at the focal point. The mirror will return parallel light rays (a beam of light) upon reflection.

On the other hand, a convex mirror does not focus parallel light rays but spreads them upon reflection. Yet even in this case, a virtual focal point can be defined for the mirror. As shown in Figure 4.8, the reflected rays appear to have a point origin behind the mirror at the focal point F.

Activities

Visible Reflections: You will need to locate a flat mirror (Plastic Mirrors, Arbor Scientific), a concentrated light source (laser, flashlight, overhead projector, slide projector), and the fog spray. Place the mirror on a table. Mount the concentrated light source above and to one side of the mirror. In a darkened room, direct the light at an angle to the surface of the mirror. Spray the fog over the mirror to make the light visible. Change the angle of the beam and repeat. Notice that the beam reflects at the same angle it hit the mirror surface—the Law of Reflection.

Figure 4.8

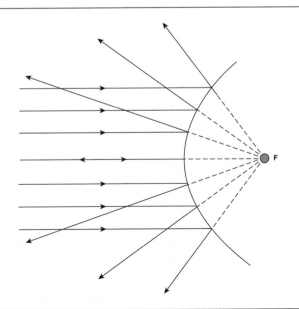

Hold additional mirrors in the path of the reflected light and notice what happens.

It is also interesting to mount a mirror on a wall and, in a darkened room, reflect a laser beam off of it. Make the beam visible with the fog spray. You might also try to reflect the laser off two or more mirrors mounted at different points in the room.

Target Practice: You will need a flat mirror and a concentrated light source. Place an object somewhere in the room and see if a student, holding a flat mirror, can reflect the light from the light source to hit the target. Have students note the orientation of the mirror for a successful hit.

Mirror and Diffuse Reflection 1: You will need a flat mirror, a concentrated light source, and salt or sand. Place the mirror on a table. In a darkened room, shine the concentrated light source at the mirror and reflect the light onto a screen, ceiling, or wall. Slowly sprinkle more and more salt or sand onto the mirror. Notice that the reflected beam becomes increasingly diffuse.

Mirror and Diffuse Reflection 2: You will need to locate two shiny metal cans and sandpaper. Shine up the two cans so that you can see your image in them. Now use the sandpaper to rough up one of the cans. Notice that your image is no longer visible.

Diffuse Reflection: You will need to locate a shoe box (with lid) and a picture (drawing or photograph). Mount the picture inside the shoe box on one side and poke a viewing hole in the shoe box through the side opposite the picture. With the lid off, look at the picture through the hole. Then try looking at the picture with the lid on. Objects are seen in reflected light (diffuse reflection). When the lid is on, there is very little light in the box and the picture is not visible.

Magic Wand: You will need a slide projector with slide, a movable screen, and a thin stick (yardstick or pointer). Darken the room and focus the slide onto a screen set a few feet in front of a doorway. Open the door and take away the screen. Move the thin stick rapidly up and down at the former position of the screen. An image of the slide will magically appear. Light cannot be seen until it reflects off something.

Multiple Reflections: You will need to locate some flat mirrors, a cup (or some other object), and a laser. The mirrors that work best for this activity are circular mirrors that come with built-in support stands. They can be purchased at almost any drugstore. More often than not, these mirrors are two-sided: one side is a flat mirror and the other a concave mirror. Place the cup (the object) on a table and see if you can arrange three (or more) mirrors (using the flat side) on the table in a way that allows you to "see" the image of the cup in the *last* mirror. Make sure all the mirrors are participating in the reflections. Draw a picture of your successful arrangement. Now hold the laser where the cup was sitting. Shine the beam into the center of the first mirror and see if it exits the last mirror. Spray the fog into the beam (in a darkened room) to see the multiple reflections. Draw a picture showing the laser beam path within your mirror system.

Minimum Mirror: You will need to locate a long mirror (the type that you attached to the back of a door in a bedroom), paper, tape, and a yardstick. Attach the mirror to a wall. Stand in front of the mirror and move until you can see an image of your whole body—the full length of your body, from your feet to the top of your head. Using the paper and tape, paper off the lower and upper parts of the mirror to find the *minimum* size mirror you need to see the whole length of your body. Measure the length of this "minimum mirror." How does it compare to your height? (Answer: Half your height) How does the distance you are in front of the mirror affect the answer? (Answer: It does not)

Mirror Reflection: You will need to locate a comb, a piece of paper, either a slide projector or an overhead projector, a concentrated light source, and mirrors (flat and curved). Tape the piece of paper to a desktop. Hold the comb, teeth down, on the paper. In a darkened room, shine the concentrated projector light through the comb, making a series of parallel light beams. You will need to adjust the angle of the light and comb to make sure that the parallel light beams are visible on the paper. Place the mirrors in the beams and note the reflections on the paper.

Light Box Reflection: You will need to purchase a light box and power supply from a science supply house (Light Box, Light Box Power Supply, Arbor Scientific). The very versatile box comes with a variety of attachments (masks, mirrors, lenses, color filters, etc.) that allow you to investigate not only reflection, but also refraction, total internal reflection, dispersion, and color mixing. You will also need to locate some white paper. Set the light box up so that it is producing some parallel light beams (see light box instructions) on the piece of paper. Place various mirrors (flat and curved) on the paper in the parallel beams. Observe and draw the reflections.

Using only one of the light beams, see if you can place multiple mirrors in its path in order to direct the beam totally around the light box, full circle.

Multiple Mirror Images: You will need to purchase three long mirrors (the type that you fasten to the back of a bedroom door). Mount two mirrors a few feet apart, with the mirror surfaces parallel and directly facing each other. Stand between the mirrors and look into one of them. Notice the multiple images, along with the "infinite hallway" or "barbershop" effect.

Now mount two of the mirrors on a table at right angles to each other, with the long side of each mirror on the table. Observe the multiple images in the mirrors. Place an object in front of the mirrors and observe. Change the angle and observe how the number of images changes.

Next, mount three mirrors in the shape of a triangle, the triangle being formed by the long sides of the mirrors. Place this triangle of mirrors on four desks, leaving an opening in the middle so that students can crawl under the desks and poke their head inside the triangle of mirrors. Notice the multiple images.

Backward Reflections: You will need to locate a flat mirror. Write a word on a piece of paper. Hold it in front of a mirror. Notice that the image of the word is backward. Now try to write a word backward in such a way that the mirror image appears forward. Try writing a word on a transparency. Hold it in front of the mirror. What does the writing in the image look like? Flip the transparency; what do you see in the mirror?

Flame Burning Underwater: You will need to locate a glass plate (an 8"-by-10" piece of glass from a picture frame works well), a piece of cardboard, a candle, and a glass of water. Place the candle between the glass plate (mounted vertically on a desktop) and the piece of cardboard (also mounted vertically on the desktop). The cardboard needs to be approximately half the height of the glass plate, tall enough to block your view of the candle, but short enough to allow you to see the candle's reflection in the glass. On the other side of the glass, place a glass of water such that the image of the candle appears submerged in the water. Darken the room and light the candle. Stand behind the cardboard shield and observe; the candle seems to be burning underwater! The glass reflects the candle light to your eye, creating an image of the candle flame behind the mirror. The glass of water can be seen at the same time because the glass is transparent and allows light from the water glass to come directly to you.

Flat, Convex, and Concave Mirrors: You will need to locate flat, convex, and concave mirrors. Convex and concave mirrors can be purchased from a science supply house (Arbor Scientific). Convex mirrors can also be found in car part stores (used by drivers to see more of the field of view than flat mirrors). Concave mirrors can be purchased from a drugstore or bathroom store (used to enlarge an image, like a face). Have students investigate image formation in all three types of mirrors (location, size, orientation, etc.). Make sure to investigate the two different kinds of images formed by a concave mirror and investigate the transition at the focal point.

Convex and Concave With Spoons: Locate a large, shiny, serving spoon. Investigate the images formed in the concave and the convex sides of the spoon. On the concave side, as you look at your image, move a finger toward the spoon and observe what happens when it crosses the focal point near the spoon.

Periscopes: You will need to locate two empty boxes of the kind used for wax paper or aluminum foil, four flat mirrors (Plastic Mirrors, Arbor Scientific), and duct tape. Make the two periscopes as shown in Figure 4.9.

Use the duct tape to tape the mirrors to the boxes. Have students view objects using both types of periscopes and note the differences in the images. You can also purchase a similar periscope from Arbor Scientific (Periscope).

Optical Illusion Toys: There are a number of toys and novelty items that use mirror reflection and image formation to produce a variety of optical illusions. They can be purchased from science supply houses (Disappearing Coin Bank, Shrinking Coin Bank, The Mirage, True Mirror; Educational

Figure 4.9

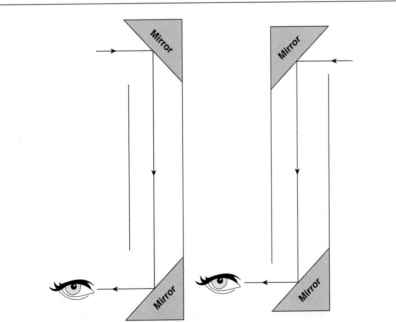

Innovations) and novelty stores. These provide intriguing ways to initiate discussion on light reflection and image formation.

Kaleidoscope: You will need to purchase three long mirrors from a science supply house (1.625"-by-9" plastic mirrors, Arbor Scientific) and some duct tape. Tape the three mirrors together to make a kaleidoscope (mirrors facing inward). Look through the kaleidoscope at various objects and note the multiple reflections.

Everyday Examples

Seeing Objects: The diffuse reflection of light, both natural and artificial, from objects in our surroundings allows us to see and locate objects.

The Moon and Planets: We see the moon and the planets because light from the sun has been diffusely reflected from them to the earth. Sometimes, on a clear night, you can see an artificial satellite in orbit about the earth as a result of the light reflection.

By the Light of the Moon: We see much better on a moonlit night than we do on a moonless night. This is an example of the double diffuse reflection. First, light from the sun is diffusely reflected off the moon and to the earth; second, this light is diffusely reflected again from objects in our surroundings to our eyes.

Day and Night: Day is bright because of the diffuse reflections of sunlight all around us. Night is dark because there is little or no light to reflect diffusely from objects around us.

One-Way Window: A person in a well-lit room at night cannot see objects outside the window, since very little light is coming into the room from the outside, but a person on the outside can see in, since the diffuse reflected light goes out the window.

Dark Clouds: Rain clouds appear dark when viewed from the surface of the earth because the tops of the clouds reflect the sunlight back up into the sky. But when flying above the clouds, you see this reflected light and the clouds appear bright white.

Bar Code Readout: A bar code reader works by directing a beam of light across the bar code (strips of black and white) and detecting the intensity of the reflected light. The black strips reflect less light than the white strips. The reader translates the pattern into numbers that are sent to a computer for identification, pricing, and inventory.

Mirror Reflectors: The following objects are, at least to some degree, mirror-type reflectors: polished metal doorknobs, polished dinnerware (like gold, silver, copper, brass, aluminum, steel), water surfaces (like puddles, ponds, lakes, swimming pools, rivers, oceans), some polished plastics, polished car and truck surfaces, polished floors, windows, mirrors, and urethane-coated wood, to name only a few.

Flat Mirrors: Flat mirrors are used to form a life-size image. They are also used to direct and redirect light beams in a variety of technological and scientific applications.

Double Reflections From Glass: You often see double reflections (and images) from light reflecting off a single pane of glass. This is a result of the fact that the light is reflecting not only off the front side of the glass pane but also reflecting off the back side as well. Furthermore, double windows (storm window or double pane) can produce four reflections (front and back of each window).

Rearview Mirrors: Rearview mirrors are used in a variety of vehicles (cars, trucks, RVs, snowmobiles, ATVs, motorcycles, bicycles, etc.) so you can see objects behind you. Bicycle riders also have headgear that comes with a small flat mirror extension that allows them to see what's behind them.

Ambulance Writing: Backward writing on the front of an ambulance appears forward when seen in a rearview mirror, as does the writing of stickers on the back window of a car.

Sunlight and Moonlight on the Water: Light from the sun or moon, when the sun or moon is near the horizon, can appear as a column of light when reflected from a relatively calm body of water (with only some small waves). The light from the sun or moon mirror-reflects off different parts of the water surface and into your eyes. On a perfectly calm body of water, a regular image (not a column) of the sun or moon can be formed.

Reflecting Telescopes: Some telescopes use concave mirrors to focus the faint light from the stars and galaxies.

Dry Versus Wet Road: Headlights work because the road diffusely reflects the light from your headlights back to your eyes. When it rains, though, much of the light mirror-reflects forward and only some of it is diffusely reflected back to your eyes. The road appears dark and you might even think that your car lights are not turned on. The headlights of cars approaching you are also mirror-reflected from the wet road and unfortunately shine straight into your eyes. Be careful driving on rainy nights!

Shiny Wet: Wet people look shiny because the water on their skin mirror-reflects some of the light.

Wet Head and Greasy Hair: Some hair sprays coat the hair with petrochemicals to make it smoother and look shinier. Greasy hair is slick and looks shiny.

Mirror Fun Houses: Curved mirrors in a fun house at an amusement park can make you look funny (shorter, taller, fatter, skinnier, etc.).

Mirror Mazes: Mazes constructed of mirrors are difficult to negotiate because of all the images and multiple reflections.

Barbershop or Infinite Hallway Illusion: When you are in a barbershop sitting between two mirrors facing each other, you can see a multiple image of yourself and what appears to be a infinitely long hallway.

Convex Mirrors in Stores and Streets: You often see convex mirrors in store aisles so that the employees can keep an eye on potential shoplifters. You sometimes find convex mirrors above street corners that help to see a larger amount of viewing space and around corners.

Face Mirrors: People use concave mirrors to enlarge the image of their face.

Periscopes: Periscopes are used at sporting events to see over crowds. Periscopes are also used in submarines to see above the water.

Kaleidoscopes: Kaleidoscopes use multiple mirrors to produce their effects.

Magic and Optical Illusions: Mirrors are often used in magic tricks and optical illusions.

Christmas Balls: Christmas tree balls act like convex mirrors.

Garden Balls: You often see large reflecting balls in yards, used as garden ornaments.

Satellite Dishes: Home and commercial satellite dishes (concave mirrors) are used to reflect and focus nonvisible TV and radio signals. Large dishes are also used by astronomers to focus visible and nonvisible light to image distant astrophysical objects.

Flashlights, Car Lights, and Lighthouses: Concave mirrors are placed behind the lightbulb in a flashlight, behind the lights in car lights, behind the lights in searchlights, and behind the lights in a lighthouse to reflect the light into a beam.

Laser Show: Moving mirrors are used in laser light shows to direct the light beams into a variety of patterns.

Water Show at Disneyland: Water fountains are used as screens to reflect cartoon images to the audience.

Radar: Nonvisible radar reflection can be used to locate objects, especially airplanes and ships. Radar is also used by police officers to detect speeding cars and by coaches to measure the speed of baseballs. Radar is used by meteorologists to determine weather patterns, including rain and wind, and by hydrologists to measure river and sewage flow. Radar is also used in some types of motion detectors.

Remote Controllers: Sometimes you can get a remote controller to work from another room by reflecting the infrared signal off a wall or door.

Mirror Signals: You can reflect sunlight off a mirror to a distant friend to signal your location.

Solar Cookers: Cylindrical concave mirrors reflect and focus light onto the food to be cooked.

Solar Farms: In some solar farms sunlight is reflected off an array of mirrors to a central collection site to generate heat and electricity.

Safety Reflectors: Reflectors are used in a variety of safety applications—road and warning signs, bike reflectors, reflectors on joggers, and more.

Eyes at Night: At night, you can often see the eyes of animals due to reflection of light (car lights, flashlights, campfires) from the moist surface of their eyes.

Ghost Images: Sometimes TV signals are received both directly and after reflecting off a nearby building. Since these two signals arrive slightly after one another, the TV picture can show weak ghost images offset from the real picture images.

Disco Ball: A disco ball that hangs over a dance floor is made up of many small flat mirrors. Color spotlights and/or laser beams reflect off the spinning ball to produce the light show.

REFRACTION

Concepts

All frequencies of electromagnetic waves travel at exactly the same speed in a vacuum, namely at 299,792,458 meters per second or about 186,000 miles per second. But this is not true for visible light when it propagates through a transparent substance like water, glass, clear plastic, or diamond. Delays in the atomic absorption and re-emission processes within the medium give rise to a net speed of propagation that is slower than the speed of light in a vacuum. Furthermore, this net speed has a slight dependence on the frequency of the light being transmitted through the substance, with higher frequencies (violet end of the spectrum) traveling at slightly shower speeds than the lower frequencies (red end of the spectrum).

Scientists often quantify the speed of light in different substances through an index number—called the *index of refraction*—assigned to the substance. This index number, when divided into the speed of light in a vacuum, gives the speed of light in the substance. For example, the index of refraction for water is 1.33, for crown glass 1.5, and for diamond 2.4. The index of refraction for air under normal temperature and pressure conditions is 1.000293, which is very close to the index of refraction for the vacuum at exactly 1.0.

The fact that the speed of light has a different value in different substances gives rise to a fascinating property of light called *refraction*. While light travels in straight lines in a given uniform substance, a light beam will bend (refract) when it propagates from one substance into another of a different index. In other words, light refracts when it either slows down or speeds up at the interface between two substances. The nature of the refraction (say, for light entering water from the air) is shown in Figure 4.10.

The amount of refraction depends on only two factors. First, the degree of bending depends on the difference of the indexes of refraction of the two substances, with larger differences giving more refraction (say, air to diamond). In other words, more refraction occurs between substances

Figure 4.10

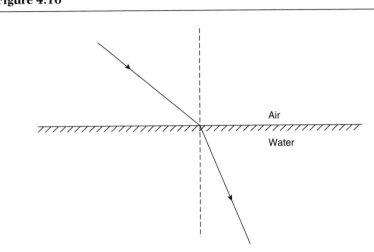

where there is a larger difference in the speed of light. Second, the degree of bending depends on the angle that the incident light ray makes with the surface between the substances, with a more grazing angle giving more refraction than a ray that is incident more perpendicularly to the surface. In fact, a ray coming straight at and exactly perpendicular to the surface will not refract at all and will enter straight into the other substance without bending. It is important to note that besides the refraction being addressed here, some of the light also reflects off the surface between the two substances according to the Law of Reflection. This reflection is not shown in Figure 4.10.

So why does light refract in this way? Part of the answer is based on the fact that light is a wave phenomenon and the other part on the fact that the speed of light (the speed of the wave propagation) changes between the two substances. In Figure 4.11a, a light beam (say, a laser beam) is shown incident at an angle to the surface between two substances (say, air to glass). The beam has been drawn showing its wave character, with the parallel lines representing the crests of the waves.

To understand the origin of the refraction, imagine each of the parallel lines as if it were the axle of a car with wheels at each end, as shown in Figure 4.11b. In analogy to the light waves slowing down as they enter the glass, assume these axles are rolling on cement (analogous to air), then entering and slowing down in mud (analogous to glass). Since one wheel will enter the mud before the wheel on the other side of the axle, this wheel will slow down first and the other will maintain its speed until it too enters the mud. This results in a turning effect for each axle. This is

Figure 4.11a

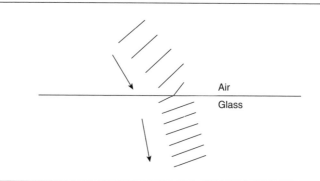

Figure 4.11b

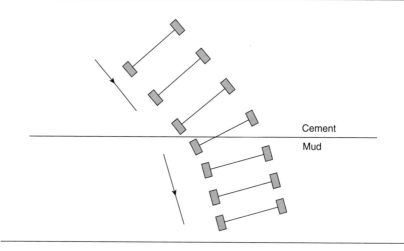

what happens for light: one end of the wavefront enters the glass first and slows down before the other side of the wavefront makes it to the glass. The light beam bends (refracts) as a result. This picture should make it clear why the amount of refraction depends on the difference of the indexes (difference in speeds) and the angle of incident. It should also make it clear why there is no refraction (although the waves change speed) when the light is directed perpendicular to the surface between the two substances.

There is a simple way to keep track of the direction of the refraction. The convention is to keep track of the refraction by referencing the bending to an imaginary line, called *the normal* (meaning "the perpendicular"), which is dawn *perpendicular* to and through the surface at the point where the ray is leaving one substance and entering the next. Figures 4.12a and 4.12b show the normals as dashed lines.

The important thing to note is that when light is incident within a substance of lower index (faster speed) and enters a substance of higher index (slower speed), say air into water, the light ray bends *toward* the normal. This is shown Figure 4.12a. When light is propagating in the opposite direction, from a higher index, slower speed substance into a lower index, faster speed substance, say water into air, the light ray bends *away* from the normal. This is shown in Figure 4.12b.

Figure 4.12a

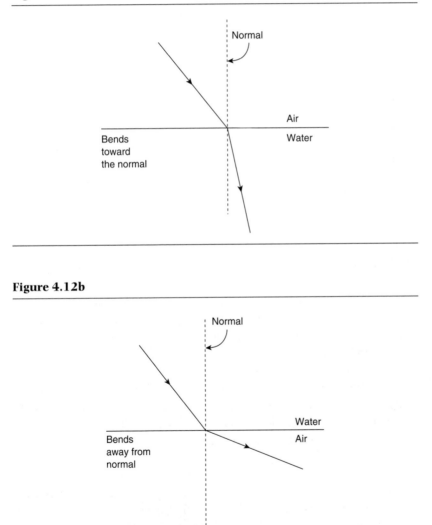

Figure 4.12b

Figure 4.13

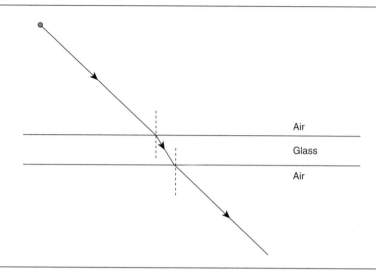

The phenomenon of refraction can be used to bend light, focus light, spread light, and create real and virtual images of various magnifications. See "Activities" and "Everyday Examples," below.

While a pane of glass refracts light, the effect is usually small. To understand this, consider the situation shown in Figure 4.13, where a ray of light is incident in air on a flat piece of glass. Ignoring all reflections, the drawing follows the two refractions, first at the air-to-glass boundary and then at the glass-to-air boundary.

In the first case, the ray bends *toward* the normal (all normals are shown as dashed lines). In the second case, the ray bends *away* from the normal. If the two surfaces are parallel to each other, like in a regular pane of glass, the ray emerges on the other side of the glass parallel to the direction it entered and displaced only slightly (unless the glass is very thick) from its original path. This is why the windows in your home do not have much of a refraction effect on what you see through them.

The same would be true of a thin layer of any transparent substance placed within another transparent substance of a different index of refraction. Still, it is fun to consider what kind of refraction would occur through a stack of thin layers of substances of increasing indexes of refraction, as shown in Figure 4.14.

Again, ignoring all reflections, the beam would bend further toward the normal each time it propagated from one substance to the next. Indeed, if the layers were all very thin, the ray would appear to curve continuously through them. This is exactly what happens when light from the sun

Figure 4.14

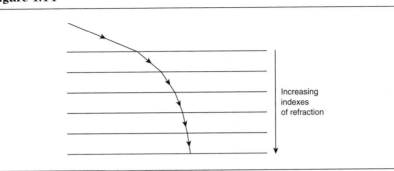

Increasing
indexes
of refraction

refracts through the atmosphere on the way to the earth's surface. Since the index of refraction decreases with elevation due to the thinning atmosphere, light rays curve downward toward the surface. See "Activities" and "Everyday Examples" for more cases of light curving.

Now consider a prism-shaped piece of glass (or water, diamond, etc.) placed in air (or in some other substance of lower index). Ignoring all reflections, Figure 4.15 follows the two refractions, first at the air-to-glass boundary and then at the glass-to-air boundary.

In the first case, the ray bends *toward* the normal (normals are shown as dashed lines). In the second case, the ray bends *away* from the normal. Unlike the pane of glass, the prism shape deflects the light ray significantly from it original path. For this reason, glass prisms are commonly used by scientists and engineers to change the direction of a light beam. For fun, consider a prism-shaped air pocket inside a chunk of glass. How would the two refractions, entering the air pocket and leaving the air pocket, be different from the glass prism case? Which way would the ray be deflected?

Figure 4.15

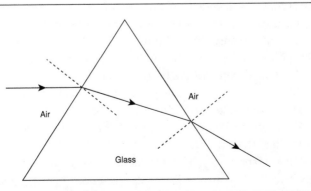

Air

Air

Air

Glass

Now consider a circular glass lens (shown edge on) surrounded by air, with one side a flat surface and the other side a convex surface (a plano-convex lens), as shown in Figure 4.16.

Figure 4.16

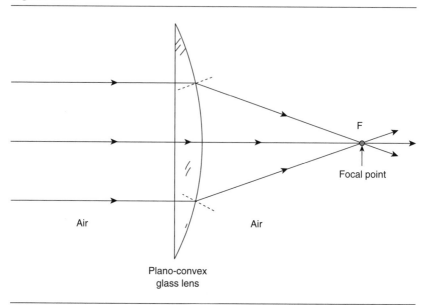

Air

Air

F

Focal point

Plano-convex
glass lens

Three parallel rays, representing a beam of light, are incident on the lens from the left. All three rays pass straight through the flat side, since there is no refraction when a light ray enters perpendicular to the surface. As the three rays emerge from the convex side, the center ray passes straight through, since it is incident directly along the normal to the glass-air interface, while the upper ray is bent downward (bending away from the normal) and the lower ray is bent upward (bends away from the normal).

This type of lens, called a plano-convex lens, is a *focusing* or *converging* lens, since the beam becomes focused to a point F, called the focal point of the lens. It makes sense that the more curved the convex side is, the closer the focal point will be to the lens. A more common type of converging lens is a double-convex lens, where both sides curve outward.

On the other hand, a *diverging* lens can be made from a circular glass lens that has one side flat and the other side concave (a plano-concave lens). As shown in Figure 4.17a, the light is not focused to a point as before, but just the opposite.

The light is spread out by the refraction; in this case, as the result of no refraction at the flat surface but with refraction (bending away from the

Figure 4.17a

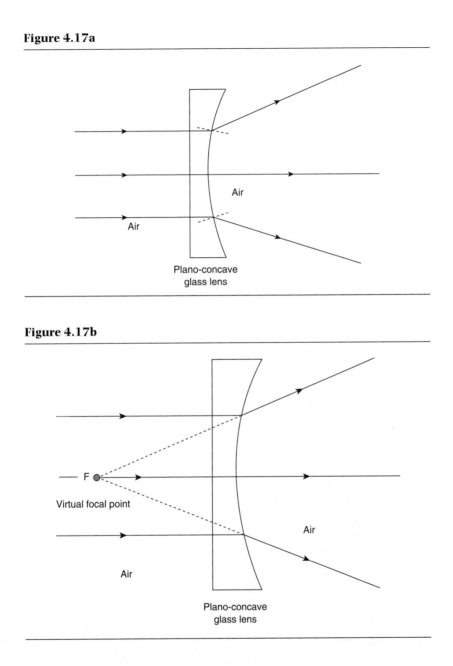

Figure 4.17b

normal) at the concave side for both the upper and lower rays. Even in this case, as shown in Figure 4.17b, a focal point F can still be defined for this lens by locating the virtual point from which the rays appear to be diverging. A double-concave lens, with both sides curving inward, is also a very common type of diverging lens.

Figure 4.18

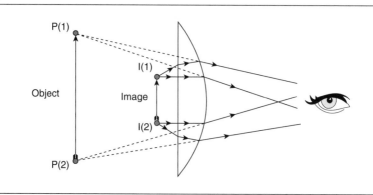

It is interesting to imagine turning a converging lens into a diverging one by simply changing the substances. For example, if the converging, plano-convex glass lens surrounded by air were replaced with a plano-convex lens of air surrounded by water or glass, the latter would be a diverging lens. Similarly, the formerly diverging, plano-concave glass lens would become a converging lens with a similar reversal of substances.

The refraction of light can be used to produce images. For example, let's consider the plano-convex lens again, but this time place an object (say, an arrow) close to one side of the lens with an observer on the other side. As shown Figure 4.18, a small cone of light emanating from two points, P(1) and P(2), representing both ends of the arrow, is refracted through the lens. As before, the eye-brain believes that the light has always been traveling in straight lines, so it "sees" these two points imaged at points I(1) and I(2). The result is a magnified, right-side-up, virtual image of the arrow.

If the arrow is placed farther from the lens, beyond the focal point, the image is much different. In this case, a real, upside-down image is formed on the side of the lens opposite the arrow. This real image can be projected onto a screen. The refractions giving rise to this image are shown in Figure 4.19.

If the image is to be projected onto a screen, the screen must be placed at the appropriate position to form a sharp picture. Most projection systems (video, slide, etc.) use a lens for projecting images onto screens. In many cases, *two* lenses are used to produce an upright image. See "Activities" and "Everyday Examples" for other applications.

A diverging lens, like the plano-concave lens analyzed earlier, can also form images. The image is virtual and smaller than the object, but right side up. This means that the lens concentrates the scene into a smaller-looking image space. For example, the peephole lens in a door, used to

Figure 4.19

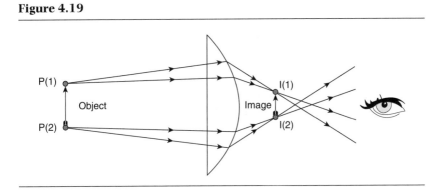

view the scene outside the door, is of this type, concentrating the larger space outside your door into a smaller image.

Activities

Aquarium Refraction: You will need to locate an aquarium or some other large clear container of water, some powdered nondairy creamer, fog spray, and two concentrated light sources (laser and flashlight). Add a couple of pinches of the dairy creamer to the water. The dairy creamer will make the light visible in the water and the fog spray will make the light visible in the air. In a darkened room, shine the laser beam through the aquarium. Spray the fog to see the light before it enters and after it leaves the aquarium. Now shine the laser from above the aquarium at the water surface. You might want to mount the laser on a tripod. Angle the beam at the surface and notice the refraction of the beam in the water. Also notice that the light may reflect off the bottom of the tank and will refract again at the surface upon leaving the water. Try different angles. Observe that the beam bends toward the normal entering the water and away from the normal when exiting. Repeat with the concentrated flashlight.

To investigate light refracting from water into the air (bends away from normal), hang the aquarium over the edge of the table and shine the laser through the bottom.

Water Refractions: You will need to locate a circular glass dish and a spherical glass flask, some powdered nondairy creamer, fog spray, and two concentrated light sources (laser and modified flashlight). Fill the dish and flask with water and add a pinch of the nondairy creamer. The creamer will make the light visible in the water and the fog spray will make the light visible in the air. In a darkened room, shine the laser beam through the side of the glass dish (parallel to the tabletop) and through the water. Use the fog spray to see the light before it enters and after it leaves the dish. Send the beam through different parts of the dish and note the refractions. A beam going straight through the center of the dish is not refracted, but one going through one side of the dish is refracted toward the other side. Observe that the light bends toward the normal when entering the water and away from the normal when exiting. Observe the focusing effect as you sweep the beam across the dish parallel to the table. Repeat with the concentrated flashlight. Now investigate the refractions thought the spherical flask.

Prism Refractions: You will need to locate a large glass prism (Giant Prism, Arbor Scientific) or a large water prism (Water Prism, Delta Education) and a concentrated light source (laser or modified flashlight). Place the laser beam a few inches above the surface of a table and shine it across the room and onto a screen. Insert the prism into the beam by standing the prism on the table. Observe the deflection of the beam due to the refractions. Dry different orientations of the prism.

Water Lens Refractions: You will need to purchase some "Refraction Cells" (Sargent-Welch) and a laser. The refraction cell is a semicircular (half-circle shape) cup that you can fill with liquid. Fill the cell with water and place it on a sheet of white paper. Place the laser on the paper and shine it through the refraction cell. Observe and draw the refractions.

Prisms and Lenses Set: You will need to purchase this set of acrylic shapes from Delta Education (Prisms and Lenses Set). The set contains the following shapes: rectangle, semicircle, triangle, double-convex, and double-concave. Set a shape on a piece of white paper and trace the path that a laser beam takes through the shape. You can also model and investigate the refraction of light through a plane of glass (rectangle), a prism (triangle), and various lenses (double-convex and -concave). For a little more money, you can purchase a similar set of shapes made from smoked glass (Smoked Lens Set, Arbor Scientific), which allows you to actually see the beam of light inside the shape.

Light Box Refractions: You will need to purchase a light box with accessories from Arbor Scientific (Light Box and Optical Set). The very versatile box comes with a variety of attachments (masks, mirrors, lenses, color filters, etc.) that allows you to investigate not only refraction, but also reflection, total internal reflection, dispersion, and color mixing. You will also need to locate sheets of white paper. Set the light box up so that it is producing some parallel light beams (see light box instructions) or a solid light beam on the piece of paper. Place various plastic shapes that come with the box—rectangle, triangle, half-circle, and lens (double-convex and -concave)—on the paper in the parallel beams or solid beam. Observe and draw the refractions.

This light box can also serve as a great light source for most of the activities described in this section.

Converging and Diverging Lenses: You will need to purchase a "Lens Set" from Arbor Scientific. The set contains a plano-convex, a double-convex, a concave-convex, and a double-concave lens. Investigate images through these lenses.

Convex Lens Image Formation: You will need a focusing lens (plano-convex or double-convex) such as those in the lens set from the previous activity. In an otherwise darkened room, allow a small amount of light to come in through part of one window. Hold the lens in that light and bring a piece of white paper slowly nearer the lens until the image (of the outside scene) is formed. Notice that the image is upside down.

You can also form an upside-down image of a lightbulb in your class-room. Remove the lampshade from a lamp in the classroom and turn on the light. Place a lens (plano-convex or double-convex) in the line of sight between the bulb and a white screen. By moving the lens toward or away from the bulb, focus the upside-down image of the lightbulb onto the screen. You might attempt to place a second convex lens in the line of sight in order to produce an *upright* image on the screen. The first lens inverts the image and the second one inverts it again.

Magnifying Water Lens: You will need a bucket and some plastic wrap. The cardboard buckets used for painting work best. Cut out a hole in the side of the bucket. The hole needs to be large enough so that your hand can be placed in the bucket. Next, place and secure a sheet of plastic wrap loosely over the top of the bucket. Pour some water on the plastic wrap. The plastic wrap should support the water in a plano-convex lens shape. While looking downward through this water lens, place your hand inside the bucket and note the magnification. Use this system to magnify various objects.

Water Drop Lens: You will need a piece of wax paper or plastic wrap and an eyedropper. Place a drop of water on the wax paper or plastic wrap. The water drop will bead up on the paper or wrap, forming a small plano-convex magnifier. Place the paper or wrap down on some writing and look through the drop. Observe the magnification. Can you use this lens to magnify other objects?

Glass of Water Magnifier: You will need a glass filled with water. Place your finger straight down through the water surface. Looking at your finger through the side of the glass (curvature of the glass forms a convex water lens), observe the magnification of the part of your finger under the water as compared to the part above the waterline. Repeat with a pencil or other objects.

Broken Pencil and Displaced Penny: You will need a pencil, a penny, two cups (one clear glass and the other opaque plastic), and water. Place the pencil inside the clear glass and set it leaning against the rim. Fill the glass about two-thirds with water and observe the pencil from the side. Note how it looks like it is "broken" at the surface. Place a penny at the bottom of the opaque glass and look down into the cup at an angle that just puts the penny out of sight. Without moving your head or the cup, pour water slowly into the cup and watch as the penny reappears.

Air Prism and Air Lens: You will need an aquarium, a large water prism (Water Prism, Delta Education), a spherical flask, a concentrated light source (laser, modified flashlight, slide projector, or overhead projector), and some powdered nondairy creamer. Take the water out of the water prism and make it an air prism. Add a pinch or two of the creamer to the water in the aquarium to make light beams visible in the water. Submerge the air prism in the water. Shine a concentrated light source through one side of the aquarium and through the air prism. Notice that the refractions are just the opposite of those for a water prism submerged in air. What happens when you submerge the same prism in water after you fill the prism with water? Now try the same thing with an air-filled spherical flask. Make sure the flask does not fill with water. Shine the light through the side of the aquarium and through the flask. Is this air lens converging or diverging? Repeat with water inside the spherical flask.

Word Trick: You will need an empty glass mayonnaise jar and a sheet of paper. Write the words CARBON DIOXIDE on the paper in one-inch-high capital letters, with the word CARBON written in red and the word DIOXIDE in blue. Take all the labels off the jar and fill it with water to overflowing and screw the lid on tightly. Place the jar sideways on the words CARBON DIOXIDE and ask for an explanation of what you see. It will appear that the letters in the word CARBON have been inverted by the water lens, yet the letters in the word DIOXIDE have not been inverted. The "trick" is that the letters in the word DIOXIDE are symmetrical letters. Each letter in DIOXIDE is indeed inverted by the lens, but the word only appears to be un-inverted. Can you think of other words with this symmetric property?

Everyday Examples

Windows: Since thin, flat glass windows refract light only a very small amount, there is very little scene distortion through a window.

Old Windows: Since some glass windows can change shape over the years and develop variation in thickness and flatness, some windows in old houses can distort the view.

Swimming Pool Light Show: When sunlight or artificial light enters a swimming pool or some other shallow body of water, the uneven surface (due to wind, etc.) can refract the light in a variety of ways. For example, areas on the water surface with convex shapes (crests) can focus the light to the bottom of the pool and form bright areas. On the other hand, surface areas with concave shapes (troughs) can spread the light out over the bottom and form darker areas. Furthermore, the light reflected from the bottom of the pool must exit through the surface again, producing additional refractions. The net effect for an observer looking into a pool is a light show of changing bright and dark areas.

Water Distortions: When you look down into a swimming pool (kids' pool, shallow pond, shallow river, shallow parts of the ocean, etc.) at objects on the bottom or patterns painted on the bottom, the objects and patterns are distorted by the surface curvatures. On the other hand, when you are submerged in water and looking up through the surface, you see dancing sunlight and distortions of the scene above.

Water Striders: Water striders are small insects that walk on the surface of many bodies of water (ponds, rivers, etc.), using the water's surface tension to support their weight on their legs. In the process, the strider's weight causes the surface to bow downward (concave) where each leg is in contact with the surface. Sunlight, in refracting through these concave areas, gets spread out. This causes a dark area of light to appear around each leg in the strider's shadow on the bottom of the pond.

Eyes and Eyeglasses: The lens of the eye is a double-convex lens used to focus light on the retina of the eye. Since the curvature of the lens and, consequently, the focal length are controlled by muscles, we have the ability to focus on objects that are different distances away. Eyeglasses are used to help the lens (and cornea) focus light onto the retina in cases where the muscles can no longer produce the required curvature for close viewing (reading glasses) and/or when the eyeball is distorted (nearsighted and farsighted). Images on the retina are upside down, but the brain reverses the images to give us an upright view. In fact, no matter what the orientation of the eye, the brain functions to keep the scene static! Turn your head side to side and notice that the scene stays where it is.

Fish-Eye View: When you are under water and looking upward at a smooth water surface, you will see the full outside scene concentrated into a circle above you.

Water Magnification: Due to refraction, objects in water, when viewed from above the water surface, appear larger than they are.

Spearing Fish and Water Depth: When you can see a fish (or anything else) underwater while looking from an angle above the water, the fish appears to be higher in the water than it actually is. Light from the fish bends away from the normal as it propagates from the water into the air and to your eye. This leads to the illusion that the fish is higher in the water than in reality. This also applies to the bottom of a swimming pool (and other bodies of water), where the bottom appears higher and, as a result, the pool shallower than it really is.

When spearing a fish from above the water, you must compensate for this effect. Also, osprey and other fishing birds usually dive straight down into the water to minimize the refraction effect.

The Archer Fish: This fish spays a thin stream of water out of its mouth to shoot down prey (mostly insects) that are above the water surface. It must learn to compensate for the refracting of light in sighting its prey. It learns to do this with practice. Young archer fish are often off target, but eventually learn to compensate for the refraction. Older archer fish often shoot from directly below their pray in order to minimize the refraction effect.

Seeing Underwater and Goggles: When you are underwater with your eyes open you see a somewhat blurry scene. This is a result of the fact that the light refracts a different amount in going from water into your eye rather than the usual case of refracting from air into your eye. This difference affects the ability of the eye to focus the light on the retina. But if you wear goggles or a swim mask, you can see clearly because the light enters the eye from air.

Shimmering Air and Water: You often see hot air rising and shimmering above a stove, hot pavement, a toaster, a campfire, a barbeque, and old home radiators. The shimmering effect results from the refraction of light through different regions of density in the rising hot air. Water being heated in a pot can show a similar shimmering effect, a result of the convection currents of varying density rising in heated water.

Twinkle, Twinkle Little Star: Stars twinkle because of the variation in the refraction of light in the earth's atmosphere. Changing air density along the path of the light causes the twinkling effect. The moon has no atmosphere, so if you were enjoying a starlit night on the moon, you would not see stars twinkling.

Upside-Down Slides in a Slide Projector: Since the double-convex lens in a slide projector reverses the image, you need to put the slides in the slide tray upside down.

Thick Glass in Aquarium Tanks: Some public aquariums have very large and deep fish tanks for viewing sea life (whales, fish, porpoises, sharks, etc.). Thick glass must be used in these tanks to resist the large water pressure. This thick glass can distort the view, especially when you are not looking straight through the glass but at an oblique angle.

Microscopes, Binoculars, and Telescopes: A series of lenses is used in microscopes, binoculars, and telescopes to magnify objects.

Magnifying Heaters: A magnifying lens can be used to focus light to start a fire or to concentrate light to heat water.

Magnifying Glass: Magnifying glasses are used by some people to read small letters. They are also used by people in a variety of professions: flytiers, jewelers, and computer engineers, to name only a few.

Reading Lens: You can buy a long, half-cylindrical lens that, when placed over a row of text, will magnify the words.

Raindrops on Leaves: Raindrops sometimes stick to leaves. A lens-shaped drop can magnify the part of the leaf under the drop. Also, the drop can focus light onto the leaf and even burn a small spot in the leaf.

Wine Glass Lens: When filled with wine, a wine glass becomes a focusing convex lens. This wine lens (plano-convex) will focus light from above the table onto the tablecloth below. A light fixture with multiple light sources, as found in most chandeliers, will produce a wonderful array of light spots on the tablecloth around the wine glass. Swirling the wine in the glass adds to the effect.

Mugs and Pitchers: The liquid contents in a thick-walled mug or pitcher are refracted by the curvature of the glass and produce the illusion that the contents fill the mug right to the edge.

TOTAL INTERNAL REFLECTION

Concepts

A fascinating phenomenon can occur when light is propagating from a higher index material (slower light speed) into a lower index material (higher light speed), for example, from water into air. As the angle of incident (measured relative to the normal in the higher index material) increases from zero degrees (perpendicular to the surface) to larger and larger angles, the proportions of the reflected and refracted light change. As the angle becomes larger, the reflected light beam gains in intensity at the expense of the refracted beam. It turns out that when a particular angle is reached, called the *critical angle*, the refracted beam's intensity becomes zero (no refraction at all) and all the light is reflected, as shown in Figure 4.20.

Also, as shown in Figure 4.20, the critical angle turns out to be the angle for which the refracted beam would have refracted through exactly 90 degrees from the normal. In other words, the refracted light does not propagate into the second material at all, but is totally reflected back into the material from which it came. This particular scenario can occur *only* for light incident within a higher index material (say, water) attempting to propagate into a lower index material (say, air) *and* when the incident angle is large enough (equal to or larger than the critical angle). It never happens in the reverse direction (air to water) or for too small an incident angle (less than the critical angle). For incident angles at or larger than the critical angle, the beam is totally reflected. The phenomenon is known as *total internal reflection.* For water to air, the critical angle is about 48 degrees, for glass to air about 43 degrees, and for diamond to air about 25 degrees.

Figure 4.20

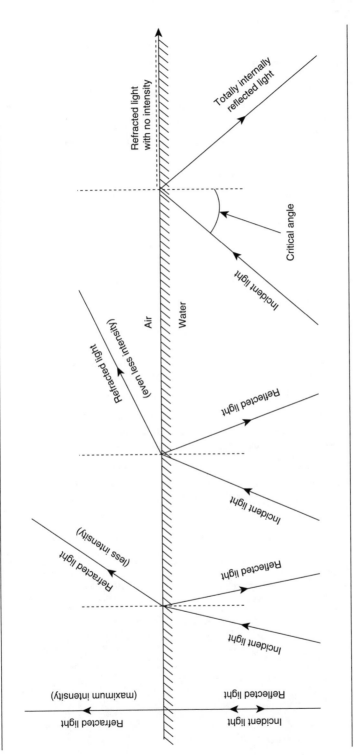

Activities

Aquarium Tank: You will need to locate an aquarium (preferably one with a glass bottom), a concentrated light source (preferably a laser), spray fog, and nondairy powdered creamer. Fill the aquarium with water and stir in a pinch or two of the powdered creamer. Slide the aquarium partway over the edge of a table or suspend it between two tables, enough so that you can shine light into the water through the bottom of the aquarium. In a darkened room, with fog spray in the air above the water surface, shine the concentrated light source through the bottom of the aquarium and toward the surface. Experiment with different angles. Find the critical angle at which you get total internal reflection from the surface. Notice that at this angle, and for larger angles, the light will not pass from the water to air, but instead reflects off the water's surface back into the water. If your aquarium does not have a glass bottom, then shine the light through one side of the tank at the surface.

Laser and Refraction Cells: You will need to locate a laser, a semicircular refraction cell (Refraction Cups, Arbor Scientific), nondairy powdered creamer, and fog spray. Fill the refraction cup with water and stir in a very small pinch of creamer. In a darkened room, with fog sprayed around the cup, shine a laser beam through the circular side of the cup and directed at the center of the flat side, as shown in Figure 4.21.

Change the angle the beam makes relative to the flat side. Notice that when the angle is large enough (measured relative to the normal), the light is totally internally reflected. You can also use the single beam from a light box instead of the laser.

Figure 4.21

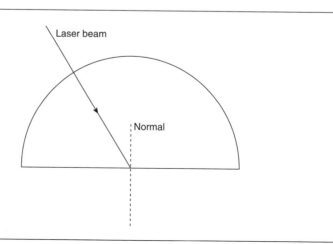

Fun With a Glass Prism: You will need to locate a right-angle glass prism (Right Angle Prism, Educational Innovations), a laser beam, and an overhead projector with screen. Shine the laser beam straight through the wide side (hypotenuse side) of the prism at one of the right-angle faces, as shown in Figure 4.22a.

The laser will strike the face at an angle of 45 degrees, which is a couple of degrees larger than the critical angle. Consequently, the beam will be totally internally reflected to the next face, where again it will be totally internally reflected, emerging from the wide side of the prism in the

Figure 4.22a

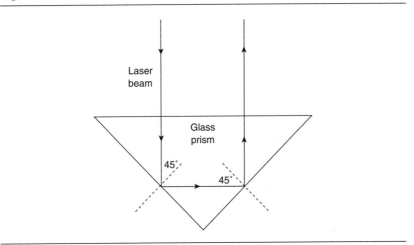

Figure 4.22b

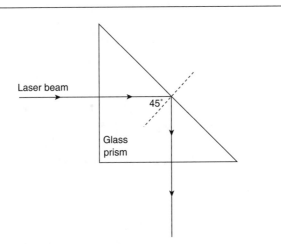

opposite direction from which it entered. You can also perform this exercise with the light box set to one or multiple parallel beams.

All the light that entered the prism is turned around as a result of total internal reflection. This result can also be observed in a darkened room when you hold the prism up in the light from an overhead projector. If the light from the overhead projector enters directly through the wide side of the prism, the light will be turned around and no light will make it through the prism to the screen. You will see the overhead projector light turned back toward the projector itself, and the prism will cast a dark shadow on a wall or screen. Also, you can place the prisms wide-side-down on the overhead and see the dark image on the screen.

You can also look straight into the wide side of a prism and see your image, since the total internal reflection returns the light back toward you, just like when you face two mirrors at right angles to each other.

You can also use total internal reflection to redirect a laser beam through 45 degrees. To do this, shine the beam straight through one of the faces at the wide side of the prism as shown in Figure 4.22b. Now, direct the overhead projector light directly through one of the faces and observe the redirected beam at 45 degrees. Also notice that a dark shadow is cast on the screen. Can you use this prism to see around a corner? Could you use two of these prisms to make a periscope?

Fiber Optics Demonstrators: You will need to purchase these wonderful total internal reflection plastic pipes (one straight and one curved) from Sargent-Welch (Optical Signal Path Demonstrator Set). You will also need a laser. The special plastic is designed so you can see the laser beam as it totally internally reflects its way through the pipe. Try other light sources.

Light Pipe: You will need to purchase this plastic spiral-shaped light pipe from Carolina Science and Math (Light Pipe). You will also need a light source. In a darkened room, shine the light into one end of the pipe and watch it being reflected (due to total internal reflection) through the spiral to the other end.

Optical Fibers: You will need to purchase an assortment of optical fibers from Carolina Science and Math (Fiber Optics Assortment). You will also need a light source. In a darkened room, shine the light into one end of the optical fiber and see it emerge at the other end. The light is totally internally reflected down the fiber.

Fiber-Optic Mineral: You will need to purchase a mineral called ulexite from Educational Innovations. Ulexite is a naturally occurring mineral composed of thousands of fibrous tubes that can transmit light due to total internal reflection. Place one side of the mineral on top of some printed words or a picture and observe the image come through to the other side.

Fiber-Optic Trees: You can purchase these fiber-optic trees at most novelty stories. Plastic fibers are bundled at the base and then spread out like the branches of a tree. A light source (often with a color wheel to continuously change the color of the light) directs light into the base of the fibers. The light is totally internally reflected though and emerges at the end of each fiber. In a darkened room, you see the light tips. Remove the fiber bundle and observe what happens when you direct light through them (laser, etc.).

 ## Everyday Examples

Endoscope and Laser Surgery: An endoscope is a flexible tube containing optical fibers that can transmit light into the body (e.g., during a colonoscopy) and return images. Laser light can be sent into the body along optical fibers to destroy cancer cells or to cauterize blood vessels.

Communications: Optical telephone cables, including high-speed data and television, are made of bundles of optical fibers that transmit digitized light signals by total internal reflection.

Polar Bear Hair: The white hairs of a polar bear act like optical fibers and transmit light to the surface of the skin to warm the skin (which, by the way, is black).

Sparkling Diamonds: Since the critical angle for diamond is only about 25 degrees, most of the light that enters a faceted diamond is totally internally reflected back and exits the way it came in, contributing to the sparkling appearance of diamond jewelry. Due to total internal reflection, the back side of a diamond in a diamond ring appears black, since very little light can make it through.

Fiber-Optic Christmas Trees: You can now buy artificial Christmas trees made with fiber-optic needles that transmit light to the needle ends.

Fiber-Optic Halloween Costumes: You can now buy Halloween costumes with built-in optical fibers that light up the costume.

Lighting: Fiber optics is used in some lighting situations where you need to pipe light to a particular location (bendable flashlights, doctors' and dentists' offices, automobile mechanics' workshops, etc.) or to keep the electric power far from the light source (swimming pool, hot tubs).

Fish-Eye View: If you are underwater and look up or if you look up from under a tank of water, you can view the results of total internal reflection. If the water surface is smooth and you are looking at the surface at an angle greater than about 48 degrees (away from the vertical), the surface looks just like a mirror. Indeed, the surface outside a cone of 96 degrees would be a mirror surface. This is a result of the light being totally internally reflected off the surface to your eye. For example, if a fish were swimming near the surface and outside the cone, you would see *two* fish. You would see the real fish plus an upside-down mirror image of the fish straight above the real one, swimming along with it.

Binoculars: Binoculars and some other optical devices use prisms to internally reflect light backward and then forward to lengthen the light path through the barrel. This turns a shorter telescope into a longer one.

Light Fountains: Some water fountains send light into the water streams. Some of the light can totally internally reflect through the streams and "follow" the streams into the fountain.

Fiber-Optic Toys: More and more light toys are being manufactured that use total internal reflection and fiber optics.

DISPERSION AND COLOR

Concepts

The continuous spectrum of visible light ranges from red (R) light at the long wavelength, low-frequency end, through orange (O), yellow (Y), green (G), blue (B), and indigo (I), up to violet (V) light at the short wavelength, high-frequency end. When the full spectrum of colors (ROYGBIV) enters the eye all at the same time, the overall color sensation is white (W). In other words, *white light* is made up of all the colors of the rainbow. The absence of all light entering the eye gives the sensation of black (Bk).

Isaac Newton first demonstrated the fact that white light is a mixture of all the colors of light. He directed a beam of white light from the sun through a glass prism. As we know, light waves refract when propagating

Figure 4.23

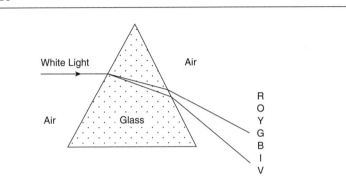

Note: R = red; O = orange; Y = yellow; G = green; B = blue; I = indigo; V = violet.

from air into glass and again when propagating from the glass back into the air. We also know that the refraction is a result of changes in propagation speeds as the light enters and exits the prism, with the amount of refraction depending on the difference in wave speeds. Since different frequencies travel at slightly different speeds in glass, with higher frequencies traveling at slower speeds, the violet light is refracted slightly more than the red light. The overall effect is to spread the white light beam out into its color components, as shown in Figure 4.23. This phenomenon is called *dispersion.*

No doubt related to the evolution of the human eye in the presence of our sun, only three colors of light (R, G, B), called the *additive primary colors,* are required to produce the sensation of white. This is probably a result of the fact that the sun produces much of its light in these three regions of the spectrum and that our eye receptors have evolved to give the sensation of red for low-frequency light in the bottom third of the spectrum, green for mid-frequencies in the middle third of the spectrum, and blue for the higher frequencies in the final third of the spectrum. Other colors can be produced from combinations of these primary colors: R + G = yellow, R + B = magenta (pinkish red), and G + B = cyan (greenish blue). Of course, if you were to combine yellow light with blue light, or magenta light with green light, or cyan light with red light, you would produce white light. For this reason, yellow and blue, magenta and green, and cyan and red are said to be *complementary colors.*

The color of pigment is a rather different story. The color of a pigment results from both the type of light incident on the pigment and which colors (wavelengths) get absorbed by the pigment and which are re-emitted. This makes pigmentation a *subtractive* process. One can make most colors of pigments by mixing magenta (pinkish red), cyan (greenish blue), and yellow pigments. Magenta pigment emits red and blue light and absorbs green light; that is, magenta subtracts green light. Cyan pigment emits blue and green and subtracts (removes) the red. Yellow pigment emits green and red and subtracts blue.

The additive primaries (R, G, B) can be produced by mixing pigments of any two combinations of cyan, magenta, and yellow. For example, mixing cyan and magenta pigments will produce a blue pigment, cyan and yellow will produce green, and magenta and yellow will produce red. Of course, mixing all three (cyan, magenta, and yellow) will produce a black or near-black pigment. Cyan, magenta, and yellow are called the *subtractive primary colors.*

Activities

Prism Dispersion: You will need to locate a prism (water or glass) and an overhead projector and screen. Mask off most of the overhead projector stage with some thin cardboard (such as the cardboard backing from a pad of paper), allowing only a narrow vertical strip of white light to hit the screen. In a darkened room, hold the prism within the strip of light and a few feet from the screen. Observe the rainbow of colors produced.

Angled Mirror in Water: You will need to locate a small mirror, an aquarium tank, and a concentrated white light source. Lean the mirror up against the sidewall, inside the aquarium tank. The mirror should lean at about a 45-degree angle. The mirror can be kept from sliding by placing a weight on the bottom of the tank against the mirror. Fill the tank with water and darken the room. Shine the concentrated white light source from above the tank into the water in such a way that it refracts (with dispersion) through the surface, reflects off the mirror, and refracts again (with dispersion) as it exits the water. Look on the ceiling of the classroom and note the rainbow of colors.

Light Box Dispersion: You will need to locate a light box (Light Box and Optical Set, Arbor Scientific), a glass prism, refraction cells, acrylic shapes, and a piece of white paper. In a darkened room, set the light box on a table on top of the paper. Put the light box into its single beam mode. Place the prism within the beam (stand the prism up vertically on the paper) and observe the dispersion. Place various refraction cells and acrylic shapes on the paper and into the beam. Investigate any dispersion.

Additive and Subtractive Color Mixing: You will need to locate three overhead projectors and color filters in the additive primaries (red, green, and blue) and in the subtractive primaries (cyan, magenta, and yellow). These filters can be purchased from Arbor Scientific (Color Filters). Place the R, G, and B filters on separate overhead projectors and project these three colors onto a single white screen. Overlap the three colors on the screen to create white light. Turn off one overhead and overlap two colors at a time to produce magenta, cyan, and yellow. Now set up the three projectors to maximize the amount of white light produced on the screen. Cast a shadow of your hand within the white light and note the "colored shadows" outside the dark shadow region. This is a result of the fact that your hand in certain locations casts a shadow of only one of the colors, allowing the other two colors to hit the screen.

Place the three subtractive primary color filters (cyan, magenta, and yellow) on a single overhead projector. Partially overlap the three filters. Observe the black region where all three overlap (all colors are subtracted), the green region for the cyan and yellow overlap (all colors but green are subtracted), the red region for the yellow and magenta overlap (all colors but red are subtracted), and the blue region for the cyan and magenta overlap (all colors but blue are subtracted).

RGB Color Pendant: You will need to purchase this crystal-shaped pendant from Arbor Scientific (RGB Color Pendant). The pendant appears white when stationary even though the three primary colors (RGB) flash at different times. When the pendant is moved quickly or whirled, the three primary colors are revealed. You can also purchase from Arbor Scientific (X-Light) a color-mixing light that fits in your pocket. It emits beams of red, green, and blue light that partially overlap to produce white light and the subtractive primaries (see "Additive and Subtractive Color Mixing," above).

Colored Shadows: You will need to purchase three clip-on desk lamps and three colored lightbulbs (red, green, blue) from your local hardware store. You will also need a white screen. Attach the three lamps to a table, forming a triangle with the lights, and project all three onto a nearby white screen. Darken the room. Place your body into the light and observe the "colored shadows" on the screen (see "Additive and Subtractive Color Mixing," above). Turn off one of the lights at a time and observe the subtractive primaries.

Color Wheel: You can buy a "magic wheel" from Educational Innovations (Magic Wheel) that is composed of many colors. When spun by the wind or a fan, the colors blend into white.

Everyday Examples

White Screens: Screens are white in order to show all colors; no colors are subtracted out.

Chromatic Aberration: Since different frequencies of visible light refract through slightly different angles, the focal point of a lens is different for different colors. This "chromatic aberration" can produce a slight ring of colors when projecting light through a lens.

Rainbows: Rainbows in the atmosphere are produced by sunlight refracting and reflecting off spherical raindrops. Rainbows can also be seen in the spray from a water hose or sprinkler. Rainbows can also be observed near a waterfall or in other situations where water spray is produced.

Beveled Mirror and Glass: White light reflecting off the edge of a beveled mirror (refraction and reflection) can be dispersed into colors. White light shining through the beveled part of a piece of glass will also show dispersion.

Faceted Pendants: A faceted glass or plastic pendant hung in a sunlit window in a room will project little rainbows on the walls, ceilings, and floors.

Jewelry and Diamond Rings: A diamond ring or other faceted jewelry will disperse white light into colors.

Moon Dogs: On some cold winter nights, colorful halos called moon dogs are formed around the moon, a result of light being dispersed by ice crystals high in the atmosphere.

Wine Dispersion: When light from a chandelier shines white light through a wine glass filled with white wine or water, colors can be seen on the tablecloth.

TV Screens: Look at a television screen with a magnifying glass. The image is made up of dots of four colors: red, blue, green, and the gray we see when the TV is off. At a distance, the close dots of red and green combine to make yellow, and so on. The gray of the screen looks black.

Paint Mixing: When you mix all colors of finger paints, watercolors, crayons, or the like, together, you get dark brown or black.

Mixing Complementary Colors: When you mix complementary colors together, all the colors are absorbed and you have black.

Pointillism: The French pointillist Georges Seurat used only magenta, cyan, yellow, and black dots to paint *Sunday Afternoon on the Island of La Grande Jatte.* From a distance, the painting appears to have more than these four colors.

Newspaper Pictures: Color photographs in a newspaper are sent and printed as the superposition of four different color transparencies. The colors are yellow, magenta, cyan, and black.

Magazine Pictures: The color illustrations in most magazines are made with magenta, cyan, yellow, and black dots.

DIFFRACTION AND INTERFERENCE

Concepts

In Chapter 3 we discussed a fundamental difference between particle motion (say, a baseball flying through the air) and wave motion (say, a sound wave propagating from a musical instrument to your ear). In particle motion matter actually moves in time through space, while in wave motion a disturbance propagates through a medium (or the vacuum) without the physical transportation of matter. Another fundamental difference between particle and wave motion is seen clearly when particles or waves interact. Particles collide with other particles and through the interaction either scatter off each other or stick together. Indeed, if the energy of the interaction is high enough, the collisions can lead to the breaking up of the particles and even to the creation of new particles. On the other hand, waves do not interact with each other in this way. Waves pass through each other and add together according to the Superposition Principle (see Chapter 3).

Another important difference between particle and wave motion, closely related to the Superposition Principle, is seen when particles and waves are forced to propagate through a narrow opening or around a small object. A fleet of boats (particles) moves straight through a breakwater opening when entering a harbor, but water waves, coming from the ocean, propagate through the breakwater and "spread out" or "bend" into the harbor. This spreading out is illustrated in Figure 4.24.

The spreading of waves when they pass through an opening or when they propagate around an object is called *diffraction*. Diffraction is a signature of a wave phenomenon. The diffraction of waves (mechanical or electromagnetic) becomes a significant effect only when the wavelength of the waves is comparable to or larger than the size of the opening or object.

Figure 4.24

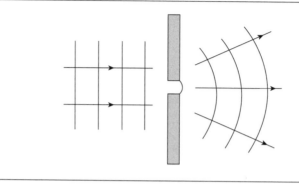

When the wavelength of the waves is very small compared to the size of the opening or object, the effect is negligible and the waves pass straight through the opening or go by the object without diffracting. This is why sound can be heard through an opening or around a corner. Audible sound waves, with wavelengths in the range from 1.7 cm for a 20,000 Hz sound to 17 m for a 20 Hz sound, come in wavelength sizes that compare to or are larger than many openings and objects, so they readily diffract. This is also why you cannot see around a corner. Visible light waves have incredibly small wavelengths ranging from 0.0000004 m for violet light to 0.0000007 m for red light. So unless you have very small openings or objects (see "Activities" and "Everyday Examples," below), visible light does not diffract much. On the other hand, larger wavelength electromagnetic waves, such as radio waves, easily diffract.

Diffraction effects were used in combination with the Superposition Principle to produce the first visible evidence that light has wave properties. Thomas Young performed some experiments in the early 1800s to try to convince the scientific community of the wave nature of light. He set up the following experiment. He sent a beam of visible light of a particular color (particular wavelength) through two small openings (he used narrow slits) close to each other (slit separation was very small). As shown in Figure 4.25, the waves diffract simultaneous through each slit. The diffracted waves from each slit cross each other and the resulting pattern is interpreted using the Superposition Principle.

If the dark lines in Figure 4.25 represent the crest of the waves and the white areas (in between the dark lines) represent the troughs of the waves, it is easy to see places where two dark lines are crossing each other—here the waves add to produce a large amplitude (double crest). There are also places in the superposition where two white areas are crossing each other—here the waves add to produce a large negative

Figure 4.25

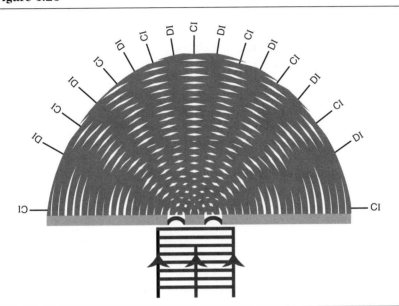

amplitude (double trough). In fact, there are a number of separated regions of these large amplitude waves in the superposition. The waves in these regions are adding up maximally together to produce regions of *constructive interference* (see regions marked CI in Figure 4.25). There are also regions where the crest of one diffracted wave is superimposed on the trough of the other (where a dark line crosses directly over a white area). In these regions the waves add up to zero and no net wave is produced—regions of *destructive interference* (see regions marked DI in Figure 4.25). For light waves, the bands of constructive interference mean large amplitudes and bright light. The bands of destructive interference mean zero amplitude and no light (black). Young demonstrated these effects by projecting the interference pattern onto a screen where bright bands of light were seen separated by dark bands. Since the amount of separation of the bright and dark bands is related directly to the wavelength of the light used (smaller wavelengths give less separation), he was able to use his experimental setup to measure the wavelengths of visible light (ROYGBIV) for the first time. When white light is used, bands of each color are produced that show up as separated rainbows on the screen.

Today, double-slit interference is no longer used to measure wavelengths of light. Instead, the resolution of the interference pattern is greatly improved by using multiple slits that are spaced very closely together (called a *diffraction grating*). See "Activities" and "Everyday Examples."

Beautiful interference colors can also be seen in thin transparent films (soap bubbles, air trapped between planes of glass, oil films, etc.). This is a direct result of the interference of light waves reflecting from the front and back surfaces of the film and superimposing to produce regions of constructive and destructive interference.

Activities

Water Wave Diffraction: You will need to locate an overhead projector and screen. You will also need to purchase a shallow, clear, flat-bottomed water tray. The clear plastic picture frame boxes used to mount photographs work very well. The tray should be large enough to cover the overhead projector's light plate. You will also need two blocks of wood and a ruler. Place the water tray on the overhead projector and fill it with about one quarter inch of water. Turn on the overhead and focus the projector until you can see waves focused on the screen. Place the two blocks of wood in the water such that a small opening (think breakwater opening in a harbor) is formed between the blocks. Direct some "flat" water waves, produced by moving the long edge of the ruler back and forth in the water, toward the opening. Observe the diffraction of the waves. Experiment with different sized openings and different wavelengths.

Light Diffraction: You will need to purchase a clear 40-watt showcase bulb with a long filament (Optical Slits Apparatus, Sargent-Welch). Turn on the lightbulb and turn off the room lights. Look at the light filament through a small slit that you form between your middle finger and pointing finger. Hold these fingers up close to one eye, oriented in the same direction as the filament. Can you see the light spread out (diffract) as it propagates through the narrow opening between your fingers? Since different wavelengths of white light diffract different amounts, you might also be able to see colors in the diffracted light.

Diffraction With Laser Light: You will need a laser with tripod, some small objects (strand of your hair, needle, etc.), a sharp-edged object (razor blade), a lens, and a screen. Mount the laser in the tripod and direct the light toward the screen. Turn off the room lights. Place a strand of hair in the laser light to cast a shadow of it on the screen. Insert a lens between the hair and the screen to enlarge the shadow. Can you see regions of interference due to the diffraction of light around the hair? Try doing the same with the point of a needle. Place the sharp edge of the razor blade in the laser beam and observe regions of interference due to the diffraction of light around the edge.

Diffraction Through Cloth: You will need a bright light source (bright bulb without the lamp shade) and a piece of cloth. Look at the light through the stretched cloth. The small holes in the cloth act like slits in a diffraction grating, so you might be able to see some interference patterns. Try different materials. Try different light sources.

Poisson's Spot: You will need black spray paint, a small piece of glass (a microscope slide works well), a laser, and a lens. Spray the paint into the air and catch some of the falling paint mist on the glass. Ideally, you will see lots of very small black dots on the glass. Let the paint dry. Shine a laser through the glass and move the glass until you cast a shadow of one of the black dots. If the conditions are right (you might have to search a number of the dots to get this to work), you might find a bright dot of light in the middle of the shadow region of one of the black dots (Poisson's Spot). You are observing light that has diffracted around the black dot and constructively interferes at the screen. You might also need to use a lens to expand the pattern.

Light Diffraction and Interference: You will need to purchase a diffraction and interference kit from Sargent-Welch (Diffraction-Interference-Resolution Kit). This kit contains equipment to produce 20 different interference patterns ranging from single-slit diffraction to double- and multiple-slit interference. It comes with an instruction manual.

Laser Interference: You will need a laser, lens, screen, and the single, double, and multiple slits from the Sargent-Welch kit described in the last activity. Shine the laser through the various slits and observe the interference patterns. If the pattern is too small to be seen easily, place a lens between the slits and the screen to expand the pattern.

Diffraction Grating Fun: You will need to purchase handheld diffracting gratings for each student in your class (Mounted Film Transmission Grating, Sargent-Welch). Hold the diffraction grating close to your eye and look at various light sources in the classroom (fluorescent, incandescent, flashlights, computer screens, candle flame, etc.) and outside the classroom (neon lights, mercury vapor lights, sodium lights, starlight, etc.) through the diffraction grating. You can also buy diffraction gratings mounted into a cardboard eyeglass frame (Prism Glasses, Educational Innovations) to view the world of interference.

Soap Bubbles and Nail Polish Remover Thin Films: You will need to purchase some soap bubble solution from a toy store and locate some nail polish remover. With the lights off and the overhead projector on, blow some bubbles into the light and observe the beautiful colors in the soap film. Place a drop of clear nail polish remover in a cup of water to form a thin film of polish remover on the surface. With the room lights off, shine a white light source at the film and observe the reflected colors.

Everyday Examples

Eyelid Diffraction: Squint your eyes and look at a bright light source. Notice the diffraction.

Umbrella Diffraction: Look up through the small holes in an umbrella on a sunny day and notice the diffraction of the light coming through the umbrella.

Spectroscopy: Scientists use diffraction gratings in concert with telescopes to detect the spectral fingerprints (light spectra) of elements in the universe. This is the primary way that scientists know the composition (elements and abundances) of the universe. Scientists and engineers use diffraction gratings and various light sources to analyze the makeup of unknown substances.

Holography: The production and viewing of holograms is based on the diffraction and interference of light. In fact, holographic techniques are used to make diffraction gratings.

Rainbows in Oil Slicks, Soap Bubble Films, Windex on Windows: You often see a rainbow of colors due to thin film interference in oily surfaces, in soap bubbles, and when washing windows with window washing fluid.

CD and DVD Colors: Colors are often seen when light is reflected from the surface of a CD or DVD. These act like reflective diffraction gratings (with the narrow grooves acting like multiple slits).

LIGHT CIRCUS

The following set of activities, selected from the activities described in this chapter, could be used to begin a unit on light. These activities would be set up around the classroom in a circus format. Next to each activity, a simple description of how each activity is to be performed would be displayed, along with a question or questions to be answered by the student in conjunction with performing the activity. Obviously, the teacher will need to rewrite these descriptions and questions to make the language and analysis appropriate for the grade level. It is suggested that students work in pairs or small groups. One option would be to have students perform the activities a few at a time and run the circus over a few days. Another option would be to use some of these activities as teacher demonstrations for whole-class discussion. In any case, students should be encouraged to probe the activities beyond the descriptions and initial questions and to begin to think of additional questions they might want to investigate on their own later in the unit.

1. Handheld Pinhole Camera

Light the candle. Hold the pinhole camera (large cup, single pinhole) between your eye and the candle flame, with the pinhole (the aluminum-foil side of the camera) toward the flame. Move the pinhole camera toward the flame until you can see an image of the flame on the screen (wax paper). What do you notice about the image? Move the camera closer to and farther away from the flame and report what you notice.

Try the pinhole camera with three pinholes. What do you observe?

Try the smaller cup. What do you observe?

2. Multiple Reflections

See if you can arrange three mirrors (or more) and the coffee cup (object) on the tabletop so that you can see the image of the coffee cup in the last mirror. Make sure all the mirrors are participating in the activity and use the flat side of each mirror (not the concave side). Draw a picture of your successful arrangement.

Now place the laser where the coffee cup was sitting. Shine the beam into the center of the first mirror and see if it exits the last mirror. Draw a picture showing the laser beam path within your mirror system.

3. Aquarium Reflection and Refraction

The aquarium tank has been filled with water. A small amount of powdered milk has been added to the water to help make visible the light beams within the water.

A. Shine the laser into the water from above the tank. Shine it at different angles to the surface and note what happens to the beam as it enters the water (observe through the side of the tank). Use fog spray to see the beam of light in the air. Draw a picture of the beam in the air and continuing its pathway in the water.

B. Now direct the laser light up through the end of the tank toward the surface of the water. What do you observe?

C. Repeat (A) and (B) with the modified flashlight white light source.

4. Shadows

Turn on the "point" light source (smaller filament). With your hand near the screen, cast a shadow of your hand onto the screen. Move your hand toward and away from the light source. What happens to the shadow?

Repeat with the "extended" light source (longer filament). What is different about this shadow compared to the shadow formed by the point source? Invent an explanation for the difference.

5. Colored Shadows

Turn on the red light (with green and blue turned off) and note the color on the white screen. Repeat with the green light (with red and blue off). Repeat with the blue light (with red and green off).

Now shine the red and green together and note the color of the mixture. Try the other combinations of two and note the colors of the mixtures.

Now shine all three together and note the color of the mixture.

With all three lights turned on, make a shadow of your hand on the screen. Describe what you observe. Can you explain your observations based on the earlier combinations you noted?

6. Fiber Optics Demonstrators, Light Pipe, and Optical Fibers

Pick up the curved plastic "light pipe" and shine the laser light into the end of the pipe. Make a sketch of what you see. Repeat with the straight plastic "light pipe." Shine the laser into the additional optical fibers provided and see if it comes out the other end.

7. Minimum Mirror

A long vertical mirror is fastened to the wall. Stand in front of the mirror and move until you can see an image of your whole body—the full length of your body—from your feet to the top of your head. Using the paper and tape, paper off some of the mirror to find the minimum size mirror you need to see your whole body. Measure the length of this "minimum mirror." How does it compare to your height? What role does the distance you are in front of the mirror play in this situation?

8. Prism and Diffraction Grating

Place the cardboard mask on the overhead projector and project a narrow line of white light onto the screen. Hold the water prism in the narrow line of light between the overhead projector and the screen and rotate the prism until a rainbow appears on the screen. The long axis of the prism should be orientated along the narrow line of light. Describe the colors. Instead of the prism, use the diffraction grating. Find the orientation that produces the clearest rainbows. Describe the colors.

9. Light Box Refractions

Turn on the light box. Shine the parallel beams of light through the different shapes (triangle, rectangle, half-circle, double-concave, and double-convex) and draw the resulting light pathways on the white paper. Describe what you noticed.

10. Converging and Diverging Lenses

Investigate images through the following lenses: plano-convex, double-convex, concave-convex, and double-concave. How do the images differ?

11. Reflections in Curved Mirrors

What do you notice when you look at your reflection in a spoon? How does it differ from looking at your reflection in a regular (flat) mirror? Try up close, far away, and in both sides of the spoon.

What do you notice when you look at your reflection in the silver Christmas ball (convex mirror)?

What do you notice when you look at your reflections in the convex and concave mirrors provided? Make sure you take a look at your image when you are close to the mirror and far away.

12. Word Trick

Look down through the plastic cylindrical lens that is sitting in a wood holder above the colored words CARBON DIOXIDE. What do you notice? Attempt an explanation.

13. Single-Slit Diffraction and Two-Slit Interference

A laser has been mounted on a small tripod and directed at a slide. The slide contains a series of double slits (two closely spaced vertical openings of different separation distances). Shine the laser light through one pair of the double slits and note the pattern on the wall.

What happens when the light goes through a single slit? Remove the double-slit slide and replace it with the other one with a series of single slits (vertical openings of different widths). Repeat. What does the pattern on the wall look like now?

SAMPLE INVESTIGABLE QUESTIONS

- *Water Wave Fun:* Can you use other clear liquids (soda, clear syrup, mineral oil, etc.) in the clear plastic tray and perform the same activities?

- *Handheld Pinhole Camera:* Can you make an "adjustable" pinhole camera with two tubes that slide over each other? How big can the hole be and still make an image?

- *Multiple Reflections:* Is it possible to set up the multiple mirrors in two different rooms and be able to see an object in one room from the other room?

- *Aquarium Reflection and Refraction:* If you had a waterproof flashlight (or sealed a regular flashlight into a watertight plastic bag) and submerged it in the water, what would you see when you directed the light straight up at the surface?

- *Shadows:* What types of light make the best shadows? Which make the worst shadows? What types of light make the largest shadow? Which make the smallest shadows?

- *Colored Shadows:* What would happen if you used different colored lights instead of red, green, and blue? What do different colored objects look like when placed in these various lights and various combinations of lights?

- *Fiber Optics Demonstrators, Light Pipe, and Optical Fibers:* What happens if you send light into the plastic pipes, light pipe, or optical fibers from both ends at the same time (say, laser light into one end and white light into the other)? Can you make a light pipe out of water?

- *Minimum Mirror:* Does the placement of the mirror on the wall (how high it is off the floor) make a difference? If so, how?

- *Prism and Diffraction Grating:* What color is a red rose when placed at various positions in the rainbow of colors on the screen?

- *Light Box Refractions:* How do things change if you use one broad beam of light instead of the parallel beams?

- *Converging and Diverging Lenses:* What happens when you shine light (modified flashlight or modified overhead projector) through these lenses?

- *Reflections in Curved Mirrors:* What would your image in a convex or concave mirror look like in a flat mirror?

- *Word Trick:* Which letters in the alphabet and which numbers (0–9) would work for this trick? Other words?

Electricity and Magnetism

ELECTRIC CHARGE AND THE ELECTRIC FORCE AND FIELD

Concepts

It is often said that atoms are the basic building blocks of nature. But, in fact, atoms are not the most fundamental entities from which our physical world is constructed. Atoms are composite structures made of even smaller and more fundamental entities. This begs the question, "What are the most basic building blocks of nature?" Indeed, what is the most basic "Lego Set of Nature" from which all objects, both animate and inanimate, are made?

To answer this very important question, let's begin by breaking apart a typical atom (typical size of 0.000000001 meters) into smaller and smaller components until we arrive at the basic Lego set. A typical atom is composed of a very small nucleus (typically 1/100,000 the size of the atom itself) surrounded by electrons in motion. By the way, atoms get their size through the motions and positions of the electrons that surround the nucleus. The *electron* is one of our basic Lego pieces. As far as we know, the electron is not composite. It is without structure. It is point-like, without size.

Although small, the nucleus is composite and can be broken down further into smaller components. The nucleus is composed of two types of particles: protons and neutrons. The proton and neutron are also composite, with size and structure, and with a radius of about 0.000000000000001 meters. They are composed of particles called *quarks*. Quarks are part of our basic Lego set. As far as we know, quarks are not composite. They are point-like, without structure and size. The proton and neutron are actually each composed of three quarks. The proton is made up of two "up quarks" and one "down quark." The neutron is made up of one "up quark" and two "down quarks." By the way, the proton and neutron get their size from the motions and positions of the constituent quarks.

Besides the *up quark* and *down quark*, there are four more flavors of quarks that make up the full family of quarks. They have been named the *charm quark*, the *strange quark*, the *top quark*, and the *bottom quark*. These six quarks are part of the basic Lego set. Since each of these quarks has its own anti-particle, the quark pieces in the Lego set total 12.

The proton and neutron are members of a much larger family of composite particles called *hadrons*. The hadron family is further subdivided into two subfamilies, *baryons* and *mesons*. Baryons are particles that are always composed of three quarks (protons and neutrons are examples of baryons) or three anti-quarks. Mesons are particles that are always composed of one quark and one anti-quark (the pion is an example of a meson). It is interesting to note that quarks, because of how they interact with each other, are found only in these combinations (three quarks, three anti-quarks, or as a quark and anti-quark pair) and *never alone* or in any other combination.

Let's get back to the electron. The electron is just one member of a larger family of fundamental particles (Lego set pieces) called *leptons*. There are six leptons (same size family as the quarks): the *electron*, the *electron neutrino*, the *muon*, the *muon neutrino*, the *tau*, and the *tau neutrino*. There is also a distinct anti-particle for each of these leptons. This gives a grand total of 12 leptons (same as the grand total of quarks).

What are the most basic building blocks of nature? *The answer is that the basic Lego set used to construct all matter in the universe is comprised of 24 pieces: six quarks, six anti-quarks, six leptons, and six anti-leptons.*

All six quarks and three of the leptons (electron, muon, and tau) have *electric charge*. The three neutrinos have no electric charge. The electron, muon, and tau each have a negative electric charge of −1 (in relative units). The anti-electron, anti-muon, and anti-tau each have a positive electric charge of +1. Indeed, charge is one (but not the only) property that distinguishes a particle from its anti-particle. The up, charm, and top quarks each have a fractional electric charge of +2/3. The down, strange, and bottom quarks each carry a fractional electric charge of −1/3. Since electric charges add, this means that the proton, composed of two up quarks and one down quark, has a net electric charge of +1 (2/3 + 2/3 − 1/3 = 1), and the neutron, composed of two down quarks and one up quark, has no net electric charge (−1/3 − 1/3 + 2/3 = 0).

Like the gravitational force that exists between objects with mass, the *electric force* exists between objects with electric charge. For example, electrons (negative charges) in an atom are attracted to and stay bound to the nucleus (positive charges) by this electric force. Indeed, besides holding atoms together, the electric force is also responsible for binding atoms and molecules together into larger structures like buildings and you. It is safe to say that the electric force is responsible for most structures and processes in our environment—excluding gravity and nuclear interactions.

While scientists do not know exactly what electric charge is, they have learned a great deal about its properties. There are only two types of electric charge, called positive electric charge and negative electric charge. Electric charges of opposite types (positive and negative) give rise to an attractive force (a pull) on each other. Electric charges of the same type (positive-positive or negative-negative) give rise to a repulsive force (a push) on each other. In simpler words, unlike charges attract and like charges repel. The strength of the attraction or repulsion grows with the amount of electric charge and decreases with increasing distance between the charges. It is interesting to note that this fact is surprisingly similar to how the gravitational force behaves between objects with mass, except in the gravitational case the force is always attractive. Gravity only pulls, while the electric force can both push and pull.

Most atoms are electrically neutral, with equal numbers of positively charged protons in the nucleus and negatively charged electrons surrounding the nucleus. In other words, most atoms have no *net* charge. Consequently, most objects made up of atoms (molecules, boats, dogs, etc.) are also electrically neutral, but this is not always the case. Electrons can be removed from an atom through various means of energy input (e.g., atom absorbs a photon of light or collides with a neighboring atom). Different types of atoms have different affinities for their electrons. Some atoms have tightly bound electrons and will give up an electron only with high energy input. Other types of atoms have loosely bound electrons and will readily give up an electron with low energy input. When an electron has been removed from an atom, the net charge of the atom—now called an *ion*—becomes +1. With two electrons missing, the net charge is +2, and so on.

Some atoms have an affinity for acquiring extra electrons. Also called an ion, an atom with one extra electron has a net charge of −1, with two extra electrons the net charge would be −2, and so on. This spread in affinities for atoms either to acquire or to give up electrons allows electric charges to be transferred from one material to another. In this way, objects can be made to acquire a net electric charge. For example, when you comb your hair, especially on a dry and low-humidity day, you will find that the comb acquires a net negative charge while your hair acquires a net positive charge, a direct result of the fact that electrons have been transferred from your hair to the comb. See "Activities" and "Everyday Examples," below, for more examples.

Metals like copper, aluminum, gold, and silver are good *conductors* of electricity. In such metals some of the electrons from the atoms are essentially free to move about inside the material. These free electrons can be made to move through the metal when pushed or pulled by other nearby charges. On the other hand, most materials—such as rubber, plastic, and wood—are good *insulators.* All of the electrons are tightly bound to their atoms and are not free to move. They can be made to conduct electricity only in extreme cases in which very large external forces can remove the tightly bound electrons from their atoms. Some special materials, called *superconductors,* become perfect conductors with no resistance to current flow at low temperatures.

It is important to note that an object with a net electric charge can attract and be attracted to a neutral object (either a conductor or an insulator). Consider the case of a conductor. If you bring a positively charged object close to a metal object, some of the electrons inside the metal will be attracted to and move through the metal toward the charged object. These electrons will end up much closer to the positively charged object. Furthermore, this process leaves behind an excess of positive charges in the metal farther from the charged object. This charge separation in the metal, induced by the nearby positively charged object, leads to a net mutual attraction, since the electrons closer to the charged object feel a larger attractive force than the smaller repulsion felt by the excess of positive charges that are farther away. This is a direct result of the fact that the electric force decreases with increasing distance between the charges. A similar situation happens when a negatively charged object is brought close to a metal object, but now some of the electrons in the metal will move away and end up much farther from the charged object, leaving behind an excess of positive charges in the metal close to the charged object. This induced charge separation leads to a net mutual attraction.

In the case of an insulator, the electrons are not free to move around in the material and do not show the same kind of charge separation as described above for the case of a conductor. A charge separation does occur, but it occurs atom by atom inside the insulator. When a positively charged object is brought near an insulator, each atom in the insulator experiences a small charge separation, with the electrons in a given atom displaced slightly toward the object and the positively charged nucleus displaced slight farther away. With this induced charge separation in each atom, each atom is slightly attracted to the positively charged object, leading to a net overall attraction of the insulator to the charged object. A negatively charged object near an insulator will also lead to a net attraction, since the atom will experience the opposite charge separation. It is interesting to note that some molecules (like the water molecule) have a built-in (non-induced) charge separation and, consequently, are readily attracted by charged objects. Such molecules are called *polar molecules*.

Sparks are often created when a charge separation has occurred, especially when the amount of charge separation is large and/or when the opposite charges are relatively close to each other. When air (or some other gas) is present in the space between the separated charges, electrons attached to the air molecules can be ripped from the molecules due to the attraction of the electrons for the external positive charges and repulsion by the external negative charges. The electrons accelerate through the air, hitting and dislodging other electrons. This cascade effect leads to electric current in the air and the emission of light (spark).

The electric force is often described in terms of the *electric field*. To understand this useful construction, first consider the gravitational force by way of analogy. A rock drops to the earth. This can be understood in terms of the gravitational force. The rock and the earth each have mass.

Since mass is the source of the gravitational force, there is a mutual gravitational attraction between the rock and the earth. But there is another way to describe the interaction between the rock and the earth. The earth's mass produces a *gravitational field* in the space surrounding the earth, and other mass (like the rock) residing in the gravitational field of the earth feels the gravitational force. Since the gravitational force on the rock decreases with increasing distance from the earth, the gravitational field must decrease in strength the farther from the earth. It is very useful (and eventually leads to a deeper understanding of all forces) to think in terms of the field concept. Mass produces a gravitational field, and other mass in the field feels the gravitational force.

The electric force can be described in a very similar and powerful way. Electric charge produces an *electric field* in the space surrounding the charge. Other electric charge residing in the electric field experiences the electric force.

It is important to note that electric charge is a *conserved* quantity. This means that the net electric charge in the universe does not change. This does not mean that electric charge cannot be created or destroyed. Negatively charged electrons, for example, are created all the time in high-energy collisions between subatomic particles, but when an electron is created in such an event, the anti-particle to the electron (called a *positron*) is also created at the exact same point in space and time. This pair creation keeps the net charge unchanged since the particle and anti-particle always have opposite charges. An electron and positron can also annihilate each other, creating two particles of light (photons) in the process. Photons do not carry electric charge, so this pair annihilation event keeps the net charge unchanged.

Activities

Important Notes:

1. The following electrostatic activities are best performed on dry, low-humidity days.

2. Most of the following electrostatic activities require the use of a negatively charged object and a positively charged object. A convenient and simple way to obtain a negatively charged object is by briskly rubbing a plastic tube (a 1-inch-diameter, 18-inch-long PVC tube from your local hardware store works well) with a piece of fur (mink or rabbit) or wool cloth. The rubbing transfers electrons from the fur or wool to the plastic tube. To obtain an object with a net positive charge, briskly rub a glass tube (same size as plastic tube, hardware store) with a piece of silk. Electrons are transferred from the glass to the silk. Make sure

the ends of the glass tube have been polished smoothed for protection. You can also buy inexpensive plastic and glass rods (with fur and silk) from Arbor Scientific (Electrostatic Demo Kit, Friction Rod Kit). Buy two kits, because you will need to have two rods of each type.

Attraction and Repulsion: You will need to buy the rods or tubes as described above. You will also need a swivel stand (Rotating Magnet Stands, Pasco). Rub the plastic rod with fur and place it on the swivel stand. Rub another plastic rod with fur and bring it up close to, but not touching, the charged rod of the stand. The rod on the stand will be repelled (pushed) by the charged rod that you are holding (like charges repel). Repeat with the two glass rods rubbed with silk. Again, like charges repel. Now place the charged plastic rod on the stand and bring the charged glass rod close to it. The plastic rod on the stand will be attracted (pulled) toward the glass rod (unlike charges attract). Notice how the electric force depends on distance. Only when the charged rods are close to each other is the electric force large enough to cause motion.

Repelling Balloons: You will need to locate two small balloons and a piece of thread. Blow up the balloons, seal them, and tie each to the end of one piece of thread (approximately 3 feet long). Suspend the balloons from the middle of the thread, so they are hanging down, side by side and touching each other. Rub the two balloons with fur or wool and let them go. Notice that the balloons now repel each other and are no longer touching. Like charges repel. You might try charging the plastic and glass rods and bringing them close to the charged balloons.

Repelling Pendulums: You will need to locate two Ping-Pong (or other table-tennis) balls, some aluminum foil, and thread. Wrap each Ping-Pong ball in foil and suspend each ball from a piece of thread from the ceiling of the classroom so that they are just touching each other. Touch a charged rod to both balls and then remove the rod. The charged pendulum balls should repel each other.

Tape Repulsion: You will need to locate some Scotch (or other cellophane) tape in a dispenser. Briskly rip off two long pieces of the tape from the dispenser. Hang one piece of tape from a finger and the other piece from a finger on your other hand. Bring the two pieces of tape close to each other and observe the repulsion. This will take a little practice because the charged tapes are attracted to your hand and do not want to hang freely.

Charge Detector: An electroscope is a very simple instrument that can be used to detect electric charge. You can purchase an inexpensive electroscope from Arbor Scientific (Electroscope, Extra Gold Foil Leaves). The electroscope is made of a metal rod that runs through a rubber stopper and into a flask. The end of the rod outside the flask has a metal ball attached. The other end of the rod, inside the flask, has a small hook over which is draped a small piece of gold foil. When a charged object is brought close to the top ball, the gold foil separates. The more charge, the more separation. Bring a positively charged rod close to (but do not touch) the ball and observe the repulsion of the gold foil in the flask. Since gold and the metal rod are good conductors, the free electrons in the metal are attracted to the positively charged rod and move into the ball. This leaves a deficiency of electrons in the gold foil (foil becomes positively charged) and a repulsion occurs (like charges repel). When the positively charged rod is removed, the electrons return and the foil collapses. Bring a negatively charged rod near the ball and observe the foil repulsion again. This time the charged rod pushes some of the electrons away from the ball and into the gold foil. The foil has an overabundance of electrons, and the like (negative) charges repel. Rub a comb through you hair and bring it close to the ball to see if you can detect the charge on the comb. Bring a balloon rubbed with fur or a piece of Scotch taped pulled from a dispenser near the ball and see if you can detect the charge. Try other objects that you think carry a net charge.

You can also cause a permanent foil separation by actually touching a charged rod (or other charged objects) to the ball.

Fun With Induced Charge Separation and Attraction: You will need the charged rods, a plastic comb, a rubber balloon, Scotch tape in a dispenser, some shredded paper, some Styrofoam pieces, salt and pepper, a pencil, and a swivel stand (Rotating Magnet Stands, Pasco). Charge up one of the rods (either the plastic rod with fur or the glass rod with silk) and bring it near some small shredded pieces of paper on a table. The paper pieces will be attracted to the rod and probably stick to the rod. If any charge is actually transferred to or from the paper pieces, they might be subsequently repelled by the rod. Try the other charged rod. Try to pick up the paper pieces using a comb that has been rubbed through your hair. Try picking up the pieces using a balloon that has been rubbed with fur or wool. Try picking up the paper with a piece of Scotch tape that has been ripped from a dispenser. Try these activities using small Styrofoam pieces or salt and pepper instead of paper shreds.

Rub the balloon with fur (or on your clothing or through your hair) and see if you can stick it to a wall or to your body.

Suspend the charged balloon from a piece of string and bring your hand slowly up to the balloon and observe the attraction of the balloon to your hand.

Place the pencil in the swivel stand. Charge up one of the rods and bring it close to one end of the pencil. The pencil will be attracted to the rod and will rotate on the stand. Repeat with the other charged rod. You might want to place other objects on the swivel stand (paper towel roll, stick of wood, etc.) to see if they will be attracted to the charged rods.

Fun With Permanent Charge Separation and Attraction: You will need the charged rods, a comb, a balloon, a water faucet, and some soap bubble solution. Charge up the plastic rod and bring it close to a slow stream of water pouring from a water faucet. Observe the attraction of the water (polar molecules) to the charged rod. Repeat with the glass rod. Repeat with a charged balloon or charged comb. Now blow some soap bubbles into the air and bring a charged rod (balloon, comb) near them. Observe the attraction.

Lots of Fun With the Van de Graaff Generator: You will need to purchase a Van de Graaff generator. This is a very versatile instrument for investigating many aspects of the electric force. A good and inexpensive one can be purchased from Arbor Scientific (Hand Crank Van de Graaff Generator). Also locate pieces of Styrofoam, tissue paper, aluminum pie tins, soap bubbles, a candle, and two helium-filled Mylar balloons. When the generator is in operation, negative electric charge is deposited on the generator's metal sphere. The generator can be discharged by touching a smaller grounded sphere to the metal sphere.

Place some pieces of Styrofoam on top of the uncharged metal sphere. Crank up the generator. Some of the charges on the sphere will be transferred to some of the Styrofoam pieces, and those pieces will jump up and off the sphere—like charges repel.

Cut up some strips of tissue paper and tape them to the uncharged metal sphere. Crank up the generator. As charge is transferred to the strips, they will be repelled by the charges on the sphere and stand on end. The Arbor Scientific generator comes with an accessory (electric plume) that shows the same effect.

Place a stack of pie tins upside down on the top of the uncharged metal sphere. Crank up the generator and watch the pie tins fly up and off the sphere. Pie tins are conductors, so electrons are transferred quickly to the pie tins.

Charge up the generator and blow some soap bubbles in the vicinity of the sphere. The soap bubbles (polar molecules) are attracted to the sphere and accelerated toward the sphere.

Bring a candle flame close to the charged sphere. Since the candle flame has a net positive charge, the flame will be attracted to the sphere.

Tie each of two helium-filled Mylar balloons to a light conducting wire (instead of string) and attach the other end of each wire to the top of the generator sphere. Crank up the generator and observed the repulsion of the balloons. Electrons are transferred from the sphere through the wires to the balloons—like charges repel.

Everyday Examples

Atoms and Molecules: The electrons in atoms are bound to the nucleus by electric forces. Atoms bind together to form molecules through electric forces.

Chemistry: Chemical reactions are electrical in nature.

Liquid Cohesion and Adhesion: Atoms and molecules in liquids cohere to each other or adhere to other substances through electrical forces.

Solids and Strength of Materials: Atoms bind together through electric forces to form solids. Solids get their structural strength from the electrical forces.

Contact Forces: Contact forces between objects, like a bat hitting a baseball, are electrical in nature.

Nuclear Fission: Some large unstable nuclei split to form two smaller nuclei. The mutual repulsion between the positive protons in the nucleus is responsible for this fission.

Hair-Raising Event: When you rub your hair with wool, like when you pull a sweater or shirt over your head, your hair can acquire a net electric charge. Some of your hair will stand on end. Like charges repel.

Party Balloon Trick: When you rub a balloon on your shirt, you can give a net charge to the balloon and it will stick to objects.

Photocopier Static: Paper that has gone through a copying machine often acquires a net charge through the copying process. The sheets of paper have a tendency to stick to together or to stick to other objects.

Electric Issues With Tape: Scotch tape or wrapping tape can acquire a net charge during the process of being unrolled from a dispenser. It will be attracted to itself or other objects before you have a chance to use it.

Electric Issues With Clothes Dryers: Socks and other clothes often stick together when rubbed together inside a clothes dryer. There are products you can buy to place in your dryer that will reduce the static buildup.

Static Cling: Clothes sometimes stick to your body because of acquired electric charge. There are spray-on products you can buy to reduce this static cling.

TV Screens and Computer Monitors: Electrons hitting the screen of a CRT (cathode ray tube) like a TV screen or computer monitor often leave a net negative charge on the screen. This is why dust tends to collect on TV screens and computer monitors. Also, a piece of paper will stick to the screen.

Lightning: A lightning flash results from a large electric charge separation between a cloud and the ground.

Sparks: You often experience a small spark when you rub your shoes over a carpet and reach out to grab a doorknob or when you slide across your car seat and reach for the door handle. Sometimes a spark can be made to jump between people.

Bedsheet Lightning: If you pull apart two bedsheets in a dark room, you sometimes see a flash of light. This is due to electric charges being separated in the pulling process.

Batteries: Batteries maintain opposite electric charges on their two terminals. The positive terminal can maintain an excess of positive charges and the negative terminal can maintain an excess of negative charges. When a simple wire or a complex electric circuit is connected between the two terminals, electrons will flow through the wire or circuit (an electric current).

Friction: The frictional force, such as the force between car tires and the road, is electrical in nature.

Electrostatic Precipitators: Electrostatic precipitators remove pollutants from waste exhaust through the electrical force.

Capacitors: Capacitors are important electronic devices that separate and store electric charges.

ELECTRIC CURRENT AND THE MAGNETIC FORCE AND FIELD

Concepts

Moving electric charge is called an *electric current.* Since most of the atomic and subatomic world is made up of particles in constant motion and most of these particles carry electric charge, it should come as no surprise that electric currents are everywhere. Electric current can be measured as the rate of charge flow. Electric current is also assigned a direction of flow, historically defined to be in the direction of the flow of positive charges. This was an unfortunate assignment, since we now know that the charge carriers in a wire are electrons with negative charge. While it is understood that electrons are the charge carriers moving in a wire, the current direction is taken to be in the exact opposite direction—as if the electrons were positive charges.

Here are a few examples of electric current:

Electrons moving through a wire

Electrons moving inside an atom

Protons moving inside the nucleus of an atom

Quarks moving inside a meson and baryon

Ions moving through a liquid

Spark jumping between your finger and a doorknob

Charged particles moving through the air in a lightning bolt

Beam of electrons inside the cathode ray tube of your TV set

Charged particles erupting inside a solar flare

A fascinating force—distinct from, but intrinsically connected to the electric force—arises between *moving* electric charges. For example, if you have two straight current-carrying wires, oriented parallel to each other, with electrons flowing in the same direction in each wire, the wires will attract (pull on) each other. The strength of this attraction grows with the magnitudes of the currents in both wires and decreases with distance between the wires. If the currents are in opposite directions, the wires will repel (push on) each other, with the strength of the repulsion growing with the magnitudes of the currents in both wires and decreasing with distance between the wires. Interestingly, if the wires are oriented perpendicular to each other, no matter the direction or magnitude of the currents in each wire, there will be no force between the wires. Other orientations give other directions and strengths to the force between the wires.

While all the details of the interaction do not need to concern us at this point in the discussion, it is very important to realize that this force is distinct from the electric force. In the cases described above, the current-carrying wires were *electrically neutral* (no net charge) and, consequently, there can be no electric interaction between the wires. The wires were neutral because the number of electrons in each wire (some attached to atoms and some contributing to the current) is identical to the number of positive charges in the wire (protons in the nuclei of the atoms that make up the wire).

The new force that arises between *moving* electric charges is called the *magnetic force.* In order for two electric charges to experience the magnetic force, both charges must be moving. While two charges do not have to be moving to interact through the electric force, both charges must be moving to experience this magnetic force. In many cases, both electric and magnetic forces are operating simultaneously between interacting electric charges. In such cases, the overall interaction is called the *electromagnetic interaction* and both the electric force and magnetic force must be determined in order to calculate the net force on a charged object.

In many cases, at both the atomic and macroscopic levels, electric charges move along circular paths. Like any two currents, these circular currents can exert magnetic forces on each other. For the sake of illustration, consider the two current loops shown in Figure 5.1a.

Figure 5.1a

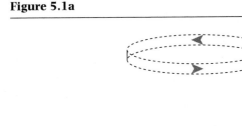

Attraction

Figure 5.1b

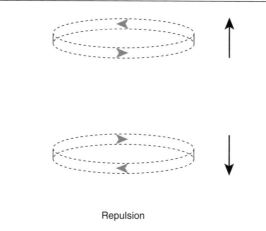

Repulsion

When the direction of current is the same in both loops and the planes of the circular loops are parallel to each other, the current loops will magnetically attract each other. This is much like the case of the mutual attraction between straight parallel wires that are carrying current in the same direction. But, as shown in Figure 5.1b, if the direction of the current is reversed in one of the loops, magnetic repulsion will result, similar to the case of parallel wires with opposite current flow.

This particular behavior might remind you of the attraction of bar magnets when the north pole of one magnet is near the south pole of the other. But, if you reverse one of the bar magnets, so that two like poles (north-north or south-south) are close to each other, a mutual repulsion results. The analogy is much more than just an analogy. Permanent magnets of any shape have unique atomic structures in which their electrons (moving around the nucleus as well as through their own intrinsic spin) can contribute circular current loops that can align themselves to produce these effects. The relationship between the attraction and repulsion of current loops and the attraction and repulsion of bar magnets is pictured in Figure 5.2.

As suggested in Figure 5.2, it is natural to assign a north and a south pole to a current loop, on each side of the loop. The convention is to assign a north pole to the side that your right thumb points to when the fingers of your right hand are curled around the loop in the direction of the current. The south pole is on the opposite side. Of course, since a permanent

Figure 5.2

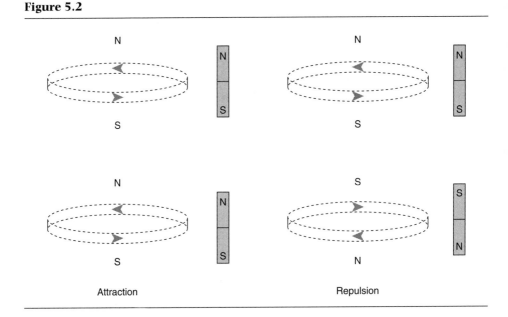

Attraction   Repulsion

magnet (like a bar magnet) is made up of aligned atomic current loops, it can attract or repel a current-carrying loop of wire, in just such a way that you would predict from the north and south pole assignments. In fact, a permanent magnet will attract or repel any electric current, the details of which depend on the strength and orientation of the magnet relative to the magnitude and direction of the electric current.

In the material that makes up a permanent magnet (naturally occurring or manufactured) there is a net permanent alignment of atomic currents that gives rise to the magnetism. In most substances, however, the atomic currents are randomly oriented with no net magnetic effect. This is why most materials are nonmagnetic. In some cases, however, a nonmagnetic material can be made magnetic by exposing it to a permanent magnet and/or an external current. This exposure can either temporarily or permanently align some of originally random atomic loops to produce a magnetic material. Iron is a common material that can be made magnetic through exposure to external magnets or currents. Indeed, a very useful magnet (often referred to as an electromagnet) can be made by looping multiple turns of wire around an iron core. When current is flowing in the wire, the magnetism is greatly enhanced by the temporarily induced magnetism in the iron core. Since the amount of current in the wire can be easily controlled, so can the strength of the electromagnet.

The important concept is that all magnetic forces result from the mutual interaction of moving electric charge. While the concept seems relatively simple, the ways in which this force can play itself out in our physical world are rich and varied. This force can manifest itself at the particle level when two moving electrons (or any two moving charged particles) exert magnetic forces on each other. It can manifest itself at the macroscopic level when currents in wires (straight wires, loops of wires, or wires in other shapes) exert pushes and pulls on each other. It can show up between two permanent magnets (made up of aligned atomic current loops) when they push or pull on each other. This force can also manifest itself between a permanent magnet and a current-carrying wire or between a permanent magnet and a moving electron (or any other charged particle). The technological applications of this force are many (see "Everyday Examples," below) and the activities are fun and engaging (see "Activities," below).

It probably comes as no surprise that the field concept can also be introduced for the magnetic force. This is done in a manner similar to how we introduced both the gravitational field and electric field. For gravity, we pictured the force between two masses as being mediated by the gravitation field. The idea was to conceive of one mass as producing a gravitational field in the space around it, and the other mass, when placed in the field, experiencing the gravitational force. For the electric force, we pictured the force between two electric charges as being mediated by the electric field. Here we conceived of one of the charges producing an electric field in the space around it, and the other charge, when placed in the field, feeling the electric force. How, then, is a field conceptualized for the magnetic force? The source of the magnetic force is moving electric charge. We can therefore introduce a new field, called the *magnetic field,* by postulating that moving electric charge produces a magnetic field in the space surrounding it, and other moving charge, when in this field, experiences the magnetic force. It is important to realize that beyond this conceptual framework, scientists have developed precise mathematical ways to determine the strength and direction of these fields.

Some closing remarks are in order. Back in Chapter 1 we argued that there really is no such thing as absolute motion. All motion is relative. So what are the implications of this fact for electric and magnetic forces? A simple thought experiment, involving two different observers (Nick and Corinne), will reveal one very important implication. One observer, Nick, is in his lab observing a charged particle at rest relative to him. He places another charged particle (also at rest relative to him) in the field of the first and records the electric force between them. As far as Nick is concerned, there is no magnetic force here because the charged particles are not moving (relative to him). But what does this exact same experiment look like to a second observer, Corinne, who is moving in a straight line with constant speed relative to Nick? From Corinne's point of view, the two charged particles at not at rest. In fact, she will observe both particles moving in parallel straight lines with constant speeds. She will also see Nick moving along with the two charged particles. This means that Corinne will record a *magnetic* force (as well as an electric force) between the moving charged particles. Indeed, we must conclude that one person's purely electric interaction can become another person's electromagnetic interaction. Obviously, there is a deep relationship between the electric and magnetic fields. As we will see later, the relationship goes even deeper than the one revealed by the Nick-and-Corinne experiment. As we will soon discuss, an electric field can be used to *produce* a magnetic field and vice versa. Furthermore, in some cases these fields can separate from and leave the charges that created them and travel through space on their own as an electromagnetic wave—as light.

Activities

Magnets Interacting With Magnets and Magnetic Materials

Playing With Magnets: You will need to collect an assortment of permanent magnets. Permanent magnets now come in a large assortment of shapes, sizes, and strengths: bar shaped, cylindrically shaped, spherically shaped, ring shaped, magnetic marbles, cow magnets, magnetic tape, magnetic wire, refrigerator magnets, and compasses, among others. A good place to purchase magnets is through Arbor Scientific or Educational Innovations. Educational Innovations sells an inexpensive starter kit with an assortment of 100 magnets (Super Mini-Neodymium Magnet Assortment). Arbor Scientific sells 50 ring magnets (Surplus Neodymium Ring Magnets). You can also purchase some natural magnets (lodestone/magnetite) from Educational Innovations (Naturally Magnetic Lodestone).

Play with the assortment of magnets. Take any two magnets and see if you can get them to both push and pull. Find and draw the position(s) of the two magnets when the mutual attraction is at a maximum. Repeat for the configurations that give maximum mutual repulsion.

Dueling Magnets: You will need to locate two bar magnets and two swivel stands. You can buy the swivel stands from Pasco (Rotating Magnet Stands). Attach (tape) one bar magnet to a swivel stand and set the stand on a table (or on an overhead projector). Bring the other bar magnet (handheld) close to the one on the swivel stand and observe the motion. Can you spin the swivel-stand magnet through constant repulsion? Can you spin it through constant attraction? Now attach both bar magnets to swivel stands and place them close together on a table (or overhead projector). Spin one of the magnets and observe the mutual effects between the two. Repeat these activities with magnets of different shapes.

The Strongest Magnet: You will need an assortment of magnets (see "Playing With Magnets," above) and a box of paper clips. Permanent magnets will attract paper clips because of the temporary magnetism induced in the paper clips. See how long a chain of paper clips a given magnet can hold up against gravity. Determine which of the magnets can hold up the longest chain. In the place of paper clips, you might want to try pins, staples, or thumbtacks.

Magnet Toys and Novelty Items: You will need to locate or buy an assortment of magnetic toys. They are available from many toy stores, novelty stores, and a number of online stores (e.g., www.officeplayground.com/magnet-toy.html, and http://my.execpc.com/~rhoadley/magtoys.htm). They range from simple children's toys to fancy executive toys. Arbor Scientific sells a few magnetic toys (Ballerina and Mirror, Seal and Ball, Magnetic Marbles, Strobe Revolution). A Google search on the key words "magnetic toys" will provide plenty of hits to help you in your search. Play with these toys. Speculate on how the toys work, considering that they are made up of magnets and that like poles repel and unlike attract.

Magnetic Structures: You will need to locate a collection of magnets of different sizes and shapes (see "Playing With Magnets," above). You might also want to purchase a Magnetic Connection Kit from Educational Innovations, containing magnetic connectors and steel spheres that can be used to build a variety of fascinating structures. Challenge your students to build structures with magnets.

Paper Clip Levitation: You will need to locate a strong magnet, a paper clip, and a piece of thread. Tie the paper clip to a piece of thread approximately 15 inches long. Tape the other end of the thread to a table. Pick up the paper clip and stretch the thread upward as far as it will go. Now bring the magnet from above and close to (but not touching) the paper clip. Carefully let the paper clip go. With a little practice you will be able to keep the paper clip levitated at the end of the thread. You might need to fix the magnet to some permanent structure instead of holding it with your hand.

More Levitation Fun: You will need to locate some ring magnets (Ceramic Ring Magnets, Educational Innovations) and a pencil. You might also want to purchase other magnetic levitation toys. Take a half-dozen ring magnets and thread them onto a pencil held vertically by its lower end. Make sure, as you thread the magnets onto the pencil, that each magnet is repelling the one above and below it (like poles repel). Can you design ways to levitate magnets of other shapes?

Magnetic Village: You will need some magnets, paper clips, poster board, and marker pens. Suspend a piece of poster board between two desks or between two piles of books. Use a handheld magnet under the poster board to pull around a paper clip (or another magnet) located on the top of the poster board. Using the marker pens, design a village scene with roads, garages, parking lots, and so forth, and "drive" your paper clip around the village.

You can use the same type of setup to surprise a classroom visitor. Hide a student under the poster board by hanging curtains around the desks. When the visitor is in the room, the student mysteriously moves some object around on the top of the desk in full view of the visitor.

Magnetic Material Hunt: You will need some magnets. Have students go around the room with handheld magnets to identify objects in the room that are magnetic (attracted to a magnet) and objects that are nonmagnetic (not attracted to a magnet). Make a chart of the classroom findings.

Magnets and Iron: Run a magnet through some beach sand to see if you can attract and remove the small pieces of iron that are present in the sand. Find items around the house that contain iron and see if they are attracted to the magnet.

Magnetic Pendulum: You will need a ring magnet, a piece of string, and an assortment of additional magnets. Make a magnetic pendulum by suspending the ring magnet from a piece of string. Have it swing just over a pile of magnets that have been placed on a table. Observe the complex motion of the pendulum as it interacts with the magnets below. Try different arrangements of the table magnets.

Compass Play: You will need to purchase some small compasses (Clear Compasses, Arbor Scientific). A compass needle is a small bar magnet that is free to rotate around a central pivot. Bring two of the compasses near each other and observe the interaction. Compare to the interaction of two bar magnets (see "Dueling Magnets," above).

A compass needle constantly interacts with the earth's magnetism and aligns with the earth's magnetic field (when not in the vicinity of other magnets). Place a bar magnet on a table and distribute a dozen compasses on the table around the magnet. Be careful not to bring the magnet too close to the compasses, since this can demagnetize or re-magnetize the needle. The compass needles will align with the magnetic field being produced by the bar magnet. This magnetic field pattern is very similar to the pattern created by the earth's magnetism. Try other magnets.

Iron Filings and Magnetic Fields: You will need to purchase some iron filings (Iron Filings, Arbor Scientific) or some less messy magnetic chips (Magnetic Chips, Arbor Scientific) and a bar magnet. Place the bar magnet on a table under a piece of stiff paper. Sprinkle the iron filings (or magnetic chips) onto the paper and draw a picture of the magnetic field formed by the filings or chips. Repeat for magnets of other shapes. You can also purchase a three-dimensional magnetic field observation box from Arbor Scientific (Magnetic Field Observation Box). Place the observation box on an overhead projector and insert the bar magnet into the center of the device. Iron filings in the liquid surrounding the magnet will align with the magnetic field, making the field visible in three dimensions. The overhead projector is used to project an image of the field for the whole class.

Magnets Interacting With Current-Carrying Wires

Magnet and Current Loop: You will need to attach a small bar magnet to a swivel stand (see "Dueling Magnets," above) and locate some insulated wire (22-gauge, 100-foot spool; Arbor Scientific), and a battery (D-cell). Wind a piece of wire a few times (10 or more) around a paper towel roll to form a multi-turn wire loop. Remove the loop from the roll. Leave a few inches of wire entering and exiting the loop. Braid these wires together. Remove a small amount of the insulation from the ends of the wire. Touch the ends of the wire to the poles of the battery to create a current in the loop. Bring the current-carrying loop close to the magnet on the stand and observe the interaction. Experiment with different orientations of the loop relative to the magnet. Which orientations produce attraction? Which produce repulsion? Try periodically disconnecting and connecting one end of the wire to the battery (to stop and start the current flow) to see if you can spin the magnet and make a motor.

Magnet and Current Loop Pendulum: You will need the materials described in "Magnet and Current Loop," above, and a strong bar magnet. Unbraid the wires in the current loop. Hang the loop as a pendulum from the edge of a table. Support the loop pendulum from the wires spaced a few inches apart and taped to the table edge. You may need to lengthen the wire entering and exiting the loop. Connect the wire leads to the battery. With current flowing in the loop, move one end of a bar magnet toward the center of the pendulum loop and observe the interaction. Can you pump up the pendulum with the magnet? Try different orientations.

Currents and Compass Needles: You will need the current loop you created in "Magnet and Current Loop," above, and a compass needle (see "Compass Play"). Bring the current-carrying loop near the compass and observe its effect on the needle. Experiment with different orientations.

Magnets and Aluminum Foil Current: You will need to locate some strong magnets (horseshoe or cow magnets work well; see "Playing With Magnets," above), aluminum foil, a six-volt battery, alligator wire leads (Mini Alligator Leads, Arbor Scientific), and a simple key switch (Single Pole, Single Throw Switch, Arbor Scientific). Cut the aluminum foil into a three-foot strip approximately one inch wide. Lay the strip on top of some magnets. Run a current through the strip by connecting the ends of the strip through the alligator wire leads to the battery. Place the key switch in the circuit so you can turn the current on and off. Press the key switch to engage the current and observe the motion of the foil. Try various orientations of the magnets and foil. Try other types of magnets.

Magnet and Lightbulb: You will need to locate a strong magnet and a clear (unfrosted) lightbulb with a long and visible filament. With the lightbulb turned on, bring the magnet close to the lightbulb until you observe the filament's vibrating. The alternating current in the lightbulb filament interacts with the magnet, causing the filament to vibrate.

Nail Electromagnets: You will need to locate two large nails, wire (same type as used in "Magnet and Current Loop," above), two batteries (D-cell), and paper clips. Wind many turns of the wire along the full length of nail. Connect the wire ends (remove the insulation at the wire ends) to the battery to create an electromagnet. Use the electromagnet to pick up paper clips or other magnetic materials. Try using this electromagnet in place of the permanent magnets in the activities described above. Create a second electromagnet identical to the first and see if you can get your pair of electromagnets to attract and repel.

Commercial Electromagnet: You can buy a very powerful electromagnet from Educational Innovations (Electromagnet) or a less powerful one from Arbor Scientific (ElectroMagnet).

Electric Motor: You can buy a very simple electric motor from Educational Innovations (World's Simplest Motor) that runs off a D-cell battery.

Current-Carrying Wires Interacting With Current-Carrying Wires

Hanging Current Loop Pendulums: You will need to construct two multi-turn loops of wire (22-gauge insulated wire; see "Magnet and Current Loop," above). Turn both loops of wire on a soup can, about 20 turns each. Remove from the can. Make sure to leave about two feet of extra wire on each end; remove the insulation from the ends. You need to hang both loops as pendulums so that they are hanging down, side by side, almost touching each other, with the planes of the loops parallel. Each loop of wire needs to be connected to a separate circuit, each with a six-volt battery and a key switch. Engage both key switches so that currents are running in the same direction in each loop—note the slight attraction of the loops. Reverse the current in one of the loops (by exchanging the leads on the battery) and note the slight repulsion that results from having currents run in opposite directions.

Hanging Aluminum Foil Wires: You will need aluminum foil, two 6-volt batteries, alligator wire leads, and two key switches (see "Magnet and Aluminum Foil Current," above).

Hang two 3-foot-long, half-inch-wide strips of aluminum foil from a support. The strips need to hang freely without touching anything. They also need to hang very close to each other with their flat sides parallel. Connect each aluminum strip separately to a circuit that includes a battery and key switch. With current flowing in the same direction in each strip, note the small attraction. Note the small repulsion when the current flows are in opposite directions.

Everyday Examples

Magnets Interacting With Magnets and Magnetic Materials

Magnetic Play: Magnets are fun to play with.

Magnetic Holders: Magnets are used to hold things in place. A short list would include refrigerator magnets, magnetic earring and jewelry clasps, magnetic knife holders, magnetic boards, magnetic name tags, magnetic hooks, magnetic tools like hammers and screwdrivers, magnetic key holders, magnetic tarp and shower curtain holders, magnetic refrigerator door seals, magnetic latches, kissing dolls, can opener lid holders, cow magnets, and oil filter magnets.

Magnetic Advertising: Many companies now give out flat magnets with their name and logos.

Toys, Games, and Magic Tricks: Many toys, games, and magic tricks use the attraction and/or repulsion of magnets to invisibly push and pull on objects.

Magnetic Salvage: Magnets can be used to find lost keys or jewelry in the grass or on the beach.

Magnetic Stirrers: Two adjacent magnets, one outside a bowl under the bottom and the other inside and at the bottom, can be used as a stirrer. The outside magnet rotates, which causes the inside magnet to do the same.

Cheap Stud Finder: A magnet can be used to find nails in drywall. Since these nails are located on the wall studs, this is a good way to locate studs in a wall.

Bogus Magnetic Bracelets and Beds: Some people claim that magnets have therapeutic powers.

Magnetic Drawing: A common art toy uses iron filings that can be positioned with a magnet to make drawings.

Earth and Other Astronomical Bodies: Like many astronomical bodies, the earth produces a magnetic field.

Magnetite: Magnetite ore is a natural magnet.

Bird Migrations: Some birds use the magnetic field of the earth to sense direction and navigate.

Credit Card Magnetic Strip: Credit cards have a magnetic strip that identifies the holder's identity and account.

Magnetic Dollars: Dollar bills are attracted to magnets because of the iron salts in the ink.

Magnets Interacting With Current-Carrying Wires

Flickering Christmas Lights: Lightbulbs can be made to flicker by placing a small permanent magnet near the lightbulb filament.

Speakers, Headphones, and Telephone Receivers: A speaker produces sound using a permanent magnet near a coil of wire. When current (carrying the sound representation) in the coil interacts with the permanent magnet attached to the speaker cone, the cone is set into vibration and produces the sound in the air.

Solenoids: There are many opening and closing devices based on the attraction of an iron rod to a coil of current-carrying wire. When current flows in the coil, the rod moves into the coil and either opens or closes a valve (or switch). Examples include car solenoids, doorbells, dishwasher valves, electric door locks, pinball machine flippers, and valves in pipes, to name only a few.

Electric Motors: Electric motors are based on the mutual interaction of magnets (permanent and electromagnets) with current-carrying wires. Examples include electric cars and buses, electric toothbrushes, CD and DVD spinners, audio and VHS tape spinners, fans, garbage disposals, electric weed trimmers, electric drills, electric saws, electric screwdrivers, dishwashers, washers and dryers, elevators, blowers, garage door openers, pumps, vibrators, electric can openers, electric clocks, windshield wipers, and electric windows, to name only a few.

ELECTROMAGNETIC INDUCTION

Concepts

We know that electric charge is the source of the electric field and that moving electric charge is the source of the magnetic field. We also know that these two fields are related through the fact that all motion is relative. But there is an even deeper connection between the electric and magnetic fields that goes beyond mere point of view. Under the right circumstances, a magnetic field can actually *create* an electric field and vice versa; an electric field can *create* a magnetic field.

Michael Faraday (1791–1867) was the first scientist to realize the specific circumstances through which a magnetic field can be used to create an electric field. He observed that when a magnet was in motion near a piece of wire, a current would begin to flow in the wire. This induced current would be created only when the magnet was in motion relative to the wire (note: the wire could be moving toward a stationary magnet to produce the same effect, but this is a moot point since all motion is relative). Further analysis reveals that the moving magnet actually creates an electric field and this new electric field is what drives the current in the wire. Since the magnet is moving, the magnetic field associated with the magnet is *changing in time* at the site of the wire. Indeed, it is the *changing magnetic field that produces the electric field,* and this induced electric field drives the current.

It is very important to realize that the magnetic field could be "changing" for a number of reasons. First, a magnet—either a permanent magnet, current in a wire, or an electromagnet—could be moving or spinning. Second, an electromagnet (i.e., a current-carrying wire wrapped around an iron core) could be in the process of being turned on or off. The switching on or off produces a magnetic field that is increasing or decreasing in time, respectively. Third, the current in a wire could be changing (increasing, decreasing, alternating). Fourth, a charged particle could be accelerating (increasing or decreasing its speed and/or turning).

James Clarke Maxwell (1831–1879) was the first to hypothesize the reverse relationship, the creation of a magnetic field by an electric field. In situations where an electric field is changing in time, a magnetic field is created. If the induced magnetic field is also changing in time, then it can produce a changing electric field. Maxwell showed that these fields can self-sustain each other and propagate together as a wave—as an electromagnetic wave, light.

Activities

Magnetic Induction With a Magnet: You will need to purchase an inexpensive meter that can measure small amounts of current (Galvanometer, Arbor Scientific). You will also need insulated wire (22-gauge, 100-foot spool, Arbor Scientific) and magnets. This activity works best with strong magnets (e.g., Cow Magnet, Educational Innovations). Make a multi-turn loop of wire by winding the insulated wire (say, 20 turns) around a coffee can. Remove the wire from the can and compress and tape the wires together. Leave a foot or so of extra wire on each end of the loop. Remove an inch of the insulation from each end. Connect the ends to the galvanometer—one end to each terminal. Notice that the meter reads zero; no current is flowing in the loop of wire. Holding the loop of wire in one hand, take the magnet in your other hand and move one pole of it quickly toward the center of the loop and observe the induced current recorded on the galvanometer. Try the other pole. Stop the magnet in the center and notice that the current drops to zero. Remove it quickly and notice that a current is induced in the opposite direction. Next, try holding the magnet steady and moving the loop of wire. Try spinning the magnet near the loop of wire. Try spinning the wire near the magnet. What happens if you move the magnet toward the *side* of the loop instead of through its center? Try loops of different sizes and numbers of turns. Try different magnets. Try different speeds.

Magnetic Induction With a Current-Carrying Wire: You will need the galvanometer and the multi-turned loop of wire from the last activity. You will also need to make a second, identical multi-turned loop of wire. Locate a six-volt battery. Connect one loop of wire to the galvanometer as before. Take the second loop of wire and connect one end of it to one terminal of the battery. Leave the other end disconnected for now. Hold the two loops of wire face-to-face and very close to each other (touching is OK). Touch the unattached end of the second loop of wire to the open battery terminal and initiate a current in the second loop. Observe the current induced in the first loop. Notice that this current is induced for only a short period of time, right after the connection, when the current is *increasing* in the second loop. A steady (constant) current in the second loop does not induce a current in the first. Now release the wire from the battery terminal and notice that a current in the first loop is induced in the opposite direction. This induced current exists for only a short period of time, during the time the current in the second loop is changing (decreasing to zero). Repeat this activity with loops of different sizes and numbers of turns. Repeat with different battery sizes. Instead of holding the loops face-to-face, repeat the activity with the faces at right angles to each other. Try other orientations. What do you observe?

Hand Generators: You will need to purchase two handheld generators (Genecon, Arbor Scientific). You can also purchase these two generators as part of a kit that includes an assortment of accessories and an activity manual (The Complete Genecon Experiment Set, Arbor Scientific). When you hold the generator in one hand and turn the crank with the other, you spin a coil of wire relative to a permanent magnet. When connected to a circuit or circuit element, this action induces a current (direct current in this case). If you crank the generator in the opposite direction, a current will be induced in the opposite direction. Attach the Genecon to a lightbulb and light up the light. Try different rotation speeds. Try to light up more than one lightbulb (in series and in parallel) with the Genecon. Attach the Genecon to a loop of wire wrapped around a compass and deflect the compass needle. In fact, go back to the "Activities" section in "Magnets Interacting With Current-Carrying Wires" and repeat some of those activities using the Genecon as your current source. Attach two Genecons together. Crank on one and watch the other one spin. The first Genecon is generating the current and the second one is using the induced current to turn a motor. A motor is a generator in reverse.

Induction Flashlights: You will need to purchase an induction flashlight (no batteries needed) from Faraday Flashlight (www.faradayflashlight.com) and a hand-powered flashlight from Arbor Scientific (Dynamo Hand-Powered

Flashlight). Both of these flashlights are powered by magnetic induction. In the first case, you shake a permanent magnet back and forth through a coil inside the flashlight to light the light. In the second case, you hand crank a lever to do the same thing. Both of these flashlights come with a clear plastic body so you can see the parts in action.

Eddy Current Damping: You will need to purchase a copper tube with two plugs (one a magnet and the other not a magnet) from Arbor Scientific (Lenz's Law Apparatus). When you hold the copper tube vertically and drop the magnet plug down through it, the moving magnet induces currents to flow in the copper tube. These induced currents (sometimes called eddy currents) in the tube cause a magnetic force on the magnet plug, which retards or dampens its motion through the tube. Of course, when you drop the nonmagnetic plug through the tube, it falls through without retardation, since it cannot induce currents to flow in the copper tube.

Eddy Current Pendulum: You will need to locate a copper plate and a strong magnet. Suspend the magnet as a pendulum bob above the copper plate. Make sure the swing of the pendulum brings the magnet bob close to the plate. Swing the pendulum and observe its motion. As the magnet swings down toward the plate, the plate experiences a changing magnetic field. This induces currents to flow in the plate (eddy currents). These eddy currents produce a magnetic field that retards (dampens) the motion of the pendulum. Compare this motion to that of a regular pendulum (a nonmagnetic bob) suspended above the plate. Try other magnets as the pendulum bob. Try other types of plates (aluminum, silver, etc.).

Everyday Examples

Power Plants: Most power plants use an energy source (coal, wood, or nuclear) to heat water to produce high-pressure steam. The steam is used to a turn a turbine (magnets and coils near each other) to produce electricity for your home.

Induction Heating: There are many commercial applications of heating by induction. In most of these cases, an alternating current is sent through a coil of wire. The resulting changing magnetic field induces currents to flow in a nearby metal. The flow of current in the metal causes the metal to heat up. A good example of this application is the induction cooktop, where coils under the cooking surface produce changing magnetic fields that induce currents to flow in a metal pot placed on the surface. These cooktops are cold to the touch, but watch out if you are wearing a ring on your finger.

Backup Generators: In places where there are no electricity sources or as backup when an electric source has failed, generators are used to produce electricity (in homes, in hospitals, when camping, in trailers, in motor homes, in RVs, at field sites, etc.). Usually powered by gasoline or propane, a coil or wire is spun relative to a magnet to produce the electricity.

Car and Bicycle Generators: When you car is in motion, the generator in your car continually recharges the battery. The engine turns the generator. Bicycle generators are attached to one wheel and use the motion of the wheel to run the generator to power lights on the bicycle.

Hybrid Cars: One fuel-saving feature in hybrid cars uses generators. As the car slows down or goes down a hill, the wheels of the car are used to turn a generator. The generator charges up the batteries that are used to accelerate the car.

Induction Flashlight: Some flashlights are now made that do not need batteries (see "Induction Flashlight" in "Activities," above).

Electromagnetic Waves: The existence of all electromagnetic waves is based on changing magnetic fields' inducing electric fields and vice versa.

Microphones: Some microphones use a magnet suspended near a coil. When the pressure variations in the sound wave hit a diaphragm that is connected to either the magnet or the coil, the movement induces currents to flow in the coil. These currents are amplified and sent to a speaker to reproduce the sound.

Transformers and Inductors: These common circuit elements use electromagnetic induction to change and control currents and voltages in electrical circuits.

Metal Detectors: Some metal detectors use electromagnetic fields that can induce currents to flow in nearby metal objects. The induced currents can be detected by the magnetic field they produce.

ELECTRIC AND MAGNETIC CIRCUS

The following set of activities, selected from the activities described in this chapter, could be used to begin a unit on electric and magnetism. These activities would be set up around the classroom in a circus format. Next to each activity, a simple description of how each activity is to be performed would be displayed, along with a question or questions to be answered by the student in conjunction with performing the activity. Obviously, the teacher will need to rewrite these descriptions and questions to make the language and analysis appropriate for the grade level. It is suggested that students work in pairs or small groups. One option would be to have students perform the activities a few at a time and run the circus over a few days. Another option would be to use some of these activities as teacher demonstrations for whole-class discussion. In any case, students should be encouraged to probe the activities beyond the descriptions and initial questions and begin to think of additional questions they might want to investigate on their own later in the unit.

1. Attraction and Repulsion

Rub the plastic rod with fur and place it on the swivel stand. The rubbing gives an electric charge to the plastic rod. Rub another plastic rod with fur and bring it up close to, but not touching, the charged rod of the stand. What happens to the rod on the stand? Repeat with the two glass rods rubbed with silk. What happens? Now place the charged plastic rod on the stand and bring the charged glass rod close to it. What happens? What can you conclude from your observations in these three cases?

2. Charge Detector

The electroscope is a very simple instrument that can be used to detect electric charge. The electroscope is made of a metal rod that runs through a rubber stopper and into a flask. The end of the rod outside the flask has a metal ball attached. The other end of the rod, inside the flask, has a small hook over which is draped a small piece of gold foil. Bring a charged rod (fur on plastic or silk on glass) close to (but do not touch) the ball and observe what happens to the gold foil in the flask. Remove the charged rod and observe what happens to the foil. Rub a comb through you hair and bring it close to the ball to see if you can detect the charge on the comb. Rub a balloon with fur or rip a piece of Scotch tape from a dispenser and bring it near the ball to see if you can detect the charge. Try other objects that you think carry an electric charge. What happens when you actually touch the charged rod (or other charged objects) to the ball and then remove the rod? Invent an explanation for your observations in the non-touching versus the touching case?

3. Fun With Induced Charge Separation and Attraction

Charge up one of the rods (either the plastic rod with fur or the glass rod with silk) and bring it near some small shredded pieces of paper on a table. What happens?

Try the other charged rod. What happens?

Try to pick up the paper pieces using a comb that has been rubbed through your hair.

Try picking up the paper pieces using a balloon that has been rubbed with fur or wool.

Try picking up the paper pieces with a piece of Scotch tape that has been ripped from a dispenser.

Try these activities using small Styrofoam pieces or salt and pepper instead of paper pieces.

Rub the balloon with fur (or on your clothing or through your hair) and see if you can stick it to a wall or to your body.

Suspend the charged balloon from a piece of string and bring your hand slowly up to the balloon and observe what happens.

Place the pencil in the swivel stand. Charge up one of the rods and bring it close to one end of the pencil. Repeat with the other charged rod.

You might want to place other objects on the swivel stand (paper towel roll, stick of wood, etc.) to see what happens when a charged rod is brought nearby.

4. Playing With Magnets

Permanent magnets now come in a large assortment of shapes, sizes, and strengths: bar shaped, cylindrically shaped, spherically shaped, ring shaped, magnetic marbles, cow magnets, magnetic tape, magnetic wire, refrigerator magnets, and compasses, among others.

Play with the assortment of magnets. Take any two magnets and see if you can get them to both push and pull. Find and draw the position(s) of the two magnets when the mutual attraction is at a maximum. Repeat for the configurations that give maximum mutual repulsion.

5. Magnet Toys and Novelty Items

Play with these toys and novelty items. Speculate on how each one works, considering that each one contains magnets and like poles repel and unlike attract.

6. Magnetic Material Hunt

Go around the room with a handheld magnet and identify objects in the room that are magnetic (attracted to a magnet) and objects that are nonmagnetic (not attracted to a magnet). Make a chart of the findings.

7. Magnet and Current Loop

Touch the ends of the wire to the poles of the battery to create a current in the loop. Bring the current-carrying loop close to the magnet on the stand and observe the interaction. Experiment with different orientations of the loop relative to the magnet. Which orientations produce attraction? Which produce repulsion? Try periodically disconnecting and connecting one end of the wire to the battery (to stop and start the current flow) to see if you can spin the magnet and make a motor.

8. Currents and Compass Needles

Touch the ends of the wire to the poles of the battery to create a current in the loop. Bring the current-carrying loop near the compass and observe its effect on the needle. Experiment with different orientations. Draw what you observe.

9. Magnetic Induction With a Magnet

A loop of wire is connected to a galvanometer—an instrument that measures current flow. Notice that the meter reads zero when there is no current flowing in the loop of wire. Try some or all of the activities described below and record your observation of the current as measured by the galvanometer in each case.

Hold the loop of wire in one hand, take the magnet in your other hand and move one pole of it quickly toward the center of the loop.

Repeat with the other pole.

Stop the magnet in the center of the loop and record the galvanometer reading when the magnet is not moving.

Remove the magnet quickly from the center of the loop and notice the galvanometer reading.

Now try holding the magnet steady and moving the loop of wire.

Try spinning the magnet near the loop of wire.

Try spinning the wire near the magnet.

What happens if you move the magnet toward the *side* of the loop instead of through its center?

Try loops of different sizes and number of turns.

Try different magnets.

Try different speeds.

10. Magnetic Induction With a Current-Carrying Wire

The first loop of wire is connected to the galvanometer—an instrument that measures current flow in the loop of wire. Take the second loop of wire and connect one end of it to one terminal of the battery. Leave the other end disconnected for now. Hold the two loops of wire face-to-face and very close to each other (touching is OK). Perform the following activities and record your observations of the current flow as measure on the galvanometer.

Touch the unattached end of the second loop of wire to the open battery terminal and initiate a current in the second loop. Observe closely the properties of the current induced in the first loop.

Now release the wire from the battery terminal and observe closely the properties of the current induced in the first loop.

Repeat this activity with loops of different sizes and number of turns.

Repeat with different battery sizes.

Instead of holding the loops face to face, repeat the activity with their faces at right angles to each other.

Try other orientations.

11. Hand Generators

When you hold the generator in one hand and turn the crank with the other, you spin a coil of wire relative to a permanent magnet. Perform the following activities and record your observations.

Attach the generator to the lightbulb and crank the generator. Try different rotation speeds.

Try to light up more than one lightbulb (in series and in parallel) with the generator.

Attach the generator to a loop of wire wrapped around a compass. Crank the generator at different speeds and observe the compass needle.

Attach two generators together. Crank on one and observe the other.

SAMPLE INVESTIGABLE QUESTIONS

- *Attraction and Repulsion:* If you get one charged rod spinning on the swivel stand, can you use another charged rod to stop it, without touching the rods together? After rubbing the rod, could you use the charged fur or silk to pick up pieces of paper? What would happen if you had two swivel stands holding two charged rods and brought the swivel stands close to each other?

- *Charge Detector:* What other objects can you charge up by rubbing and then detect the charge with the electroscope?

- *Fun With Induced Charge Separation and Attraction:* What will happen if you place other objects (wood, metal rod, etc.) on the swivel stand and bring a charged rod nearby? What kinds of objects work best for charge separation and attraction?

- *Playing With Magnets:* How can you order this assortment of magnets from strongest to weakest? Could you see how long a chain of paper clips each magnet can hold to determine the relative strengths? Or maybe you could use an identical pair of magnets to see how many pieces of papers you can put between the magnets before the magnetism will no longer hold them together.

- *Magnet Toys and Novelty Items:* Can you make your own magnetic toy or game?

- *Magnetic Material Hunt:* What if you extended the hunt to outside of the classroom, to objects in the natural environment or on the playground?

- *Magnet and Current Loop:* What would happen if you used more than one battery (or a larger battery) and/or a larger or smaller current loop and/or more turns of wire? Could you use the generator (Genecon, Arbor Scientific) instead of the battery to observe the same effects?

- *Currents and Compass Needles:* What happens to the needle if you reverse the direction of current in the loop? How close does the compass need to be in order to be deflected by the current loop? What would happen if you used more than one battery (or a larger battery) and/or a larger or smaller current loop and/or more turns of wire? Could you use the generator (Genecon, Arbor Scientific) instead of the battery to observe the same effects?

- *Magnetic Induction With a Magnet:* What kinds of magnets work best for induction? Could you use an electromagnet (ElectroMagnet, Arbor Scientific) instead of the permanent magnet to produce an even larger induced current?

- *Magnetic Induction With a Current-Carrying Wire:* What would happen if you replaced the second loop with an electromagnet (ElectroMagnet, Arbor Scientific)?

- *Hand Generators:* If you connect one of the generators to a battery, will the generator become a motor? How many lightbulbs (in series) can the generator power? With the two generators connected together, can you crank on one of them and have the other one do work (move an object, etc.)? If you connect (wire) the two generators together over a long distance (on the playground), will cranking on one still turn the other? If you connect the generator to a loop of wire, can you generate enough current in the loop to heat up the wire?

References
and Resources

BOOKS

Berry, D. A. (1987). *A potpourri of physics teaching ideas.* College Park, MD: American Association of Physics Teachers.

Carlson, M., Humphrey, G., & Reinhardt, K. (2003). *Weaving science inquiry and continuous assessment.* Thousand Oaks, CA: Corwin Press.

Connect Magazine, Synergy Learning, PO Box 60, Brattleboro, VT 05753, 800–769–6199, http://www.synergylearning.org

Cunningham, J., & Herr, N. (1994). *Hands-on physics activities with real-life applications.* San Francisco: Jossey-Bass.

DiSperzio, M. (1999). *Awesome experiments in light and sound.* New York: Sterling Publishing.

Doherty, P., & Rathjen, D. (1991). *The Exploratorium science sourcebook.* San Francisco: The Exploratorium.

Edge, R. D. (1981). *String and sticky tape experiments.* College Park, MD: American Association of Physics Teachers.

Ehrlich, R. (1990). *Turning the world inside out and 174 other simple physics demonstrations.* Princeton, NJ: Princeton University Press.

Freier, G. D., & Anderson, F. J. (1981). *A demonstration handbook for physics.* Stony Brook, NY: American Association of Physics Teachers.

Gunstone, R., & Watts, M. (1985). Force and motion. In R. Driver (Ed.), *Children's ideas in science* (chap. 5). Maidenhead, Berkshire, UK: Open University Press.

Harlen, W. (2001). *Primary science, taking the plunge.* Portsmouth, NH: Heinemann.

Herbert, D. (1980). *Mr. Wizard's supermarket science.* New York: Random House.

Hewitt, P. (2006). *Conceptual physics.* San Francisco: Benjamin Cummings.

Levenson, E. (1985). *Teaching children about science: Ideas and activities every teacher and parent can use.* New York: Prentice Hall.

Liem, T. L. (1989). *Invitations to science inquiry.* El Cajon, CA: Science Inquiry Enterprises.

McCullough, J., & McCullough, R. (2000). *The role of toys in teaching physics.* College Park, MD: American Association of Physics Teachers.

National Research Council. (1999). *National science education standards.* Washington, DC: National Academy Press.

Prigo, R. (1994). Teaching physics at a liberal arts college: Creativity, appreciation, and delight. In K. W. Prichard & R. M. Sawyer (Eds.), *Handbook of college teaching: Theory and application* (pp. 281–293). Westport, CT: Greenwood.

Rossing, T. D., & Chiaverina, C. J. (2001). *Teaching light and color.* College Park, MD: American Association of Physics Teachers.

Sarquis, J., Hogue, L., Sarquis, M., & Woodward, L. (1997). *Investigating solids, liquids, and gases with toys.* Middletown, OH: Terrific Science Press.

Saul, W., & Reardon, J. (1996). *Beyond the science kit.* Portsmouth, NH: Heinemann.

Van Cleave, J. (1985). *Teaching the fun of physics.* New York: Prentice Hall.

Van Cleave, J. (1991). *Physics for every kid.* New York: John Wiley.

SCIENCE PROGRAMS AND KITS

Delta Science Modules (DSM), http://www.deltaeducation.com
Education Development Center (EDC, Insights), http://main.edc.org/
Full Option Science System (FOSS), http://www.lawrencehallofscience.org/foss/
Great Explorations in Math and Science (GEMS), http://www.lhsgems.org
National Science Resources Center (NSRC), Science, Technology and Children (STC), http://www.nsrconline.org/
Science Curriculum Improvement Study (SCIS 3+), http://www.deltaeducation.com
TOPS Learning Systems, http://www.topscience.org

SCIENCE SUPPLY VENDORS

Arbor Scientific, 800–367–6695, http://www.arborsci.com
Carolina Science and Math, 800–334–5551, http://www.carolina.com
Delta Education, 800–442–5444, http://www.deltaeducation.com
Educational Innovations, Inc., 888–912–7474, http://www.teachersource.com
Frey Scientific, 800–225–FREY, http://www.freyscientific.com
Nasco Science, 800–558–9595, http://www.eNasco.com
Pasco, 800–772–8700, http://www.pasco.com
Sargent-Welch, 800–727–4368, http://www.sargentwelch.com

Index